An Alternative Lorentz Invariant Relativistic Wave Equation

A New Tool For Analyzing The Wave Nature of Matter

By

Shelton W. Riggs, Jr.

An Alternative Lorentz Invariant Wave Equation

A New Tool For Analyzing The Wave Nature of Matter

Copyright © 2007 by Shelton W. Riggs, Jr

All rights reserved. No part of this book may be used or reproduced by any means, graphic, electronic, or mechanical, including photocopying, recording, taping or by any information storage retrieval system without the written permission of the publisher except in the case of brief quotations embodied in critical articles and reviews.

ISBN-13: 978-1-44954-373-0

ISBN-10: 1-44954-373-1

This book is dedicated to
my dear friends
Al and Nancy Kellner,
Edward and Joyce Sommer,
Burns and Valora Taylor,
Mario LaFragola,
John and Son Brian Shapiro,
Dennis and Pat Bechtlofft,
Burt Almon,
Wayne Wilson
and all others.

About The Author

Shelton W. Riggs, Jr. earned undergraduate (University of Texas) and graduate (Vanderbilt) degrees in both Physics and Mathematics.

Professionally, he has consulted as both a hardware and software design engineer to numerous Fortune 500 companies for a wide range of scientific applications. He helped solve several scientific problems for US Army, Air Force and Navy.

Other interests include theoretical physics including quantum mechanics, relativistic mechanics and theoretical mathematics (especially the mystery of prime numbers).

Hobbies include dancing, karaoke, juggling, playing keyboards, writing songs, and writing poetry.

Other Works By Author

The Scientific Theory of God – A bridge Between Faith and Physics provides the reader with basic scientific understanding, interpretation, clarification and answers about concepts and beliefs associated with a Supreme Being. These ideas are developed and based on current theory and the standard model of physics. This new basis has revealed surprising relationships between the scientific definitions of both God and man. A model for the behavior of living matter (bioenergy) has been extended to include the behavior of human beings in terms of perception, decision and action. These concepts combined with the operation of short and long-term memory explain both human consciousness and how the mind controls the body. This model also includes how any desired behavior (provided it does not go against survival) may be achieved. This book offers a scientific creation theory and shows how it is compatible with both the big bang as well as evolutionary theory.

Nature of the First Cause – The Discovery of What Triggered the Big Bang contains the formal scientific theory of how the universe got started. It lays down the mathematical foundation for the creation theory put forth in "The Scientific Theory of God" book. It resolves the asymmetry problem of physics. It solves the two main cosmological problems by identifying both dark energy and dark matter. This theory predicts the correct order of magnitude for the number of galaxies and stars in the universe revealed by the Hubble ultra deep field results. It uncovers two entangled parallel worlds consisting of negative antimatter and positive matter. It explains the accelerated expansion of both matter and negative antimatter. It predicts the distance between matter and negative antimatter to be the Schwarzschild diameter of the expanding universe.

The Origin of the Planck Mass, Planck Length and Planck Time presents solutions of a system composed of two identical photons which are trapped in each other's gravitational field. The solution applies to any pair of identical particles having zero rest mass. Two solutions were derived. One solution was found by treating two photons as point particles. The quantum mechanical solution came about by treating the two photons as waves. In both solutions, the predicted distance between photons was found to be proportional to the Planck length. The period of the photon's orbit was proportional to the Planck time and the mass energy of each photon is proportional to the Planck mass. The concept of a Planck length, Planck mass and Planck time all emerge from this single model.

Primal Proofs offer several proofs that deal with prime numbers. A proof by contradiction of Goldbach's binary conjecture that every even natural number greater than two (2) can be expressed as the sum of two (2) primes is given. A proof of Goldbach's ternary conjecture that all natural numbers greater than five (5) are the sum of three (3) primes via the binary proof is presented. A proof by construction (utilizing the proof of the binary conjecture) of the twin prime conjecture is offered. A proof of the Riemann hypothesis by deduction is presented. A proof that any prime greater than three (3) is the mean of two other primes is presented. A proof is offered that any even number greater than twelve (12) satisfies Goldbach's binary conjecture in a plurality of ways. Two entangled formulas that generate all the primes beyond the second prime ($P_2 = 3$) are developed and summarized.

Acknowledgments

I acknowledge God for providing all the resources necessary to develop a more general relativistic wave theory.

I acknowledge all my teachers especially my science and mathematics teachers.

I acknowledge every author in the reference section of this book for providing both ideas and data.

I acknowledge my parents for providing a secure and nurturing environment that initially made my learning fun.

I acknowledge my country for providing me with my freedoms, especially my freedom of speech.

Preface

This study was undertaken to investigate the possibility of obtaining a more general alternate Lorentz invariant relativistic wave equation other than the Klein–Gordon wave equation or the Dirac wave equation.

Neither the Klein-Gordon or Dirac equation can be used to describe particles with zero rest mass which move under the influence of forces described by a potential energy function.

Another disadvantage to both the Klein-Gordon equation and the Dirac equation is that both are time dependent. Thus, neither equation can be readily used (without assumptions of periodic time dependence) to describe stationary energy states inherent in bound systems which move under the influence of force fields described by a potential energy function.

Moreover, if any particle having non-zero rest mass is stationary, the corresponding Klein-Gordon or Dirac wave functions do not collapse. In contrast, the non-relativistic Schrödinger wave functions for a free particle always collapse when the particle is at rest.

This manuscript offers an invariant form which differs from both the Dirac equation as well as the Klein-Gordon equation. This generalized relativistic wave equation may be used to describe particles with zero rest mass as well as particles with non-zero rest mass. This equation is capable of describing systems moving under the influence of forces derivable from a potential energy function.

The generalized relativistic wave equation offered by the author adheres to Schrödinger's philosophy in that if a free system is at rest, then it is a particle with a collapsed wave function. In other words, motion is the source of the wave nature of matter.

The fundamental physical laws, basic units, physical constants, and basic elementary particles have been included for reference after the glossary.

Please refer to the glossary for the definitions, values and symbols for the various physical quantities within this manuscript.

Table of Contents

Leading Pages	i-ii
Title Page	iii
Copyright Page	iv
Dedication	v
About the Author	vi
Other Works By Author	vii
Acknowledgments	xi
Preface	xii
Table of Contents	xv
Chapter 1 The Klein-Gordon Equation	1
Brief History of Basic Wave Equations	1

Klein-Gordon Weaknesses 2

Incomplete Hamiltonian 3

Factors of Klein-Gordon Equation 4

Inconsistent Free Particle Wave Equation 5

Particles With Zero Rest Mass
Elude Klein-Gordon Equation 8

Chapter 2 The Dirac Equation 10

Particles With Zero Rest Mass
Elude Dirac Equation 12

Chapter 3 The Schrödinger Equation 13

Free Particle Schrödinger Wave Function 14

Origin of Schrödinger's Equations 16

Time Independent Equation 17

Time Dependent Equation 19

Table of Contents   xvii

Operator Relationships   20

Chapter 4 Relativistic Preliminaries   23

Relativistic Action Tool   24

Chapter 5 Free Particle Derivation   26

Classic Momentum Operator   27

Free Particle Time Independent Wave Equation   28

Relativistic Operators for Free Particles   29

Invariant Forms   30

The Hamiltonian   30

Chapter 6 Time Independent Equation   32

Operator Representation   32

Chapter 7 Towards the Time Dependent Equation 35

Kinetic Energy Operator 38

Chapter 8 Time Dependent Equation 40

Generalized Wave Equations 41

Three Forms for the Relativistic Hamiltonian 42

Summary of Hamiltonians 45

Chapter 9 Evolution of the Wave Function 46

Wave Function Analysis of a Particle at Rest 48

Wave Function Analysis of a Particle in Motion 51

Chapter 10 Conservation of Probability 53

Conservation of
Relativistic Probability Current 54

Rate of Change of Expectation Values 56

Chapter 11 Quantum Mechanical Implications 59

Statistical Conservation of Energy 59

Persistence of Normalization 60

Chapter 12 Summary and Conclusions 62

Time Independent Equation 62

Reduction to Schrödingers
Time Independent Equation 63

Time Dependent Equations 63

Derived Equations Summary 65

Hamiltonian Summary 66

Kinetic Energy Operator Summary 66

Conclusions	67
Evolution of the Wave Function	68
Glossary	70
Fundamental Physical Laws	73
Preliminary Definitions	73
Mechanical Laws	80
Electromechanical Laws	92
Conservation Laws	98
Basic Units	101
Equivalent Units	101
Basic Physical Constants	102
Basic Elementary Particles	104

Preliminary Particle Descriptors	104
Anti–Particle Properties	107
Matter Energy Particles	108
Field Energy Particles	111
References	118
Index	124
Trailing Pages	141 – 142

Chapter 1
Klein-Gordon Equation

Brief History of Basic Wave Equations

There exists only a total of three basic quantum mechanical wave equations. These are the non–relativistic Schrödinger wave equation, the Klein-Gordon relativistic wave equation and the Dirac relativistic wave equation.

The Schrödinger wave equation is famous for its success in describing and predicting the behavior of electrons associated with the protons of atomic nuclei.

The Klein–Gordon equation was the first Lorentz invariant relativistic wave equation to be proposed.

An Alternative Relativistic Wave Equation

The Dirac equation is famous for both its successes in describing the relativistic behavior of electrons influenced by electromagnetic fields and its prediction of the anti-electron or positron. Dirac began his development with the Klein–Gordon equation.

Klein-Gordon Weaknesses

We will begin by showing some of the inconsistencies encountered by an analysis of the Klein-Gordon relativistic wave equation. This equation was the first proposed Lorentz invariant relativistic wave equation and is written as

$$(N1)\ (\nabla^2 - \partial^2/(\partial(c^2 t^2)))\Psi(r,t) = (m_0^2 c^2/\hbar^2)\Psi(r,t)$$
$$= (E_0^2/(\hbar^2 c^2))\Psi(r,t)$$

where r is the particle's (x,y,z) position, t is the time, Ψ is the wave function, c is the speed of light in vacuum, m_0 is the rest mass of the particle (an energy system) and $\hbar$ is Planck's rationalized constant ($h/2\pi$) where h is Planck's constant. Here ∇ is the normal three dimensional Del vector operator defined as

(N2) $\nabla = (\partial/\partial x, \partial/\partial y, \partial/\partial z)$

E_0 refers to the rest mass energy m_0c^2. As shown in the Quantum Mechanical Laws section, let's try to put the equation in a form like $\hat{H}\Psi = E_T\Psi$ where E_T is the total energy eigenvalue and $\hat{H}$ is the total energy operator or Hamiltonian operator. Equation (N1) may be rewritten in a similar form as

(N1.2) $(\hbar^2/m_0)(\nabla^2 - \partial^2/(\partial(c^2t^2)))\Psi(r,t) = E_0\Psi(r,t)$

Incomplete Hamiltonian

Note that the eigenvalue turns out to be the rest mass energy, E_0 and is not the total energy. This may be a sign that there may be some missing pieces to this equation. To interpret equation (N1.2), on the left side of the equal sign, the factors to the left of the wave function namely, $(\hbar^2/m_0)(\nabla^2 - \partial^2/(\partial(c^2t^2)))$ must be the part of the Hamiltonian operator that generates the rest mass energy of a free (no unbalanced forces acting) system as its eigenvalue. The rest mass energy is actually potential energy (energy of mass at rest). Knowing that by definition the total energy is the sum of kinetic and potential energy, then the part of the Hamiltonian that is missing, must represent the kinetic energy operator. Thus, the Klein-Gordon equation represents only half of an equation which must describe free systems in motion (i.e. kinetic energy).

Factors of Klein-Gordon Equation

If one utilizes the gradient operator

(N3) $\square = (\partial/\partial x, \partial/\partial y, \partial/\partial z, \partial/\partial ict)$

to define the Dalembertian operator

$$(N3.1)\ \Box \bullet \Box = \Box^2 = \nabla^2 - \partial^2/(c^2 \partial t^2)$$

where $\bullet$ is the normal vector dot product, the Klein-Gordon equation

$$(N1)\ (\nabla^2 - \partial^2/(\partial(c^2 t^2)))\Psi(r,t) = (m_0^2 c^2/\hbar^2)\Psi(r,t)$$

can be more compactly written as

$$(N1.3)\ \Box^2\Psi(r,t) - (m_0^2 c^2/\hbar^2)\Psi(r,t) = 0$$

which can be factored as

$$(N1.4)\ [\Box - m_0 c/\hbar][\Box + m_0 c/\hbar]\Psi(r,t) = 0$$

since $m_0 c/\hbar$ is a constant.

Inconsistent Free Particle Wave Function

The corresponding free particle wave function that satisfies the Klein-Gordon equation is

(N4) $\Psi(r,t) = \Psi_0 e^{(rp - Et)i/\hbar}$

where

(N4.1) $E = mc^2$

is the total dynamic energy, and

(N4.2) $p = mv$

is the momentum, r is the system's (or particle's) three dimensional position, (x,y,z). Ψ_0 is the initial wave function and e is the base of the natural log whose imaginary exponent represents the normal oscillatory part of the wave function. Note that when the system is at rest, it's energy becomes the rest mass energy, via (N4.1) $E = mc^2$, thus

(N5) $E = E_0 = m_0c^2$

and its momentum via (N4.2) $p = mv$ is

(N6) $p = mv = 0$

since its velocity, v is zero. Plugging all this into equation

(N4) $\Psi(r,t) = \Psi_0 e^{(rp - Et)i/\hbar}$

yields

(N4.3) $\Psi(0,t) = \Psi_0 e^{-(E_0 t)i/\hbar}$

Clearly, this is inconsistent since it says that a free system at rest and having non-zero rest mass has an oscillating time dependent wave function.

This is not good philosophy since by definition, a free system at rest cannot be time dependent.

Note also that the wave function of equation

(N4) $\Psi(r,t) = \Psi_0 e^{(rp - Et)i/\hbar}$

does not contain a kinetic energy term, since $E = mc^2$ represents the total energy as a sum of both the kinetic energy $(m - m_0)c^2$ and the rest mass potential energy m_0c^2.

Particles With Zero Rest Mass Elude Klein-Gordon Equation

The Klein-Gordon equation

(N1) $(\nabla^2 - \partial^2/(\partial(c^2t^2)))\Psi(r,t) = (m_0^2c^2/\hbar^2)\Psi(r,t)$

and it equivalent,

(N1.3) $\Box^2\Psi(r,t) - (m_0^2 c^2/\hbar^2)\Psi(r,t) = 0$

are known to be Lorentz invariant, however they cannot be applied to particles with zero rest mass (such as photons, gravitons or possibly neutrinos) which move under the influence of forces derivable from a potential energy function V. For zero rest mass systems, equation (N1.3) reduces to

(N1.01) $\Box^2\Psi(r,t) = 0$

It would be difficult for such a potential to enter into any kind of generalization of equation (N1.01).

Let us now turn to a brief discussion of the famous Dirac wave equation. Recall that Dirac developed his wave equation from the Klein-Gordon equation.

Chapter 2
The Dirac Equation

As previously shown, the other way of expressing the Klein-Gordon wave equation

(N1) $(\nabla^2 - \partial^2/(\partial(c^2t^2)))\Psi(r,t) = (m_0^2 c^2/\hbar^2)\Psi(r,t)$

by its factors is equation

(N1.4) $[\Box - m_0 c/\hbar][\Box + m_0 c/\hbar]\Psi(r,t) = 0$

which is similar to how Paul Dirac factored it into the spinor notation of equation

(N7) $[\Sigma\gamma_\mu \, \partial/\partial x_\mu + m_0 c/\hbar][\Sigma\gamma_\mu \, \partial/\partial x_\mu - m_0 c/\hbar]\Psi_\mu = 0$

where, the subscript μ in the sum, Σ ranges from 1 to 4 and corresponds to x, y, z, and ict. All other symbols are defined exactly as for the Klein-Gordon equation. The γ_μ are 4 X 4 complex matrices and the wave function, Ψ_μ is a four component column vector.

Note that $\gamma_\mu \gamma_\nu + \gamma_\nu \gamma_\mu = 2\delta_{\mu\nu}$ Here, if $\mu = \nu$, then $\delta_{\mu\nu} = 1$ otherwise, $\delta_{\mu\nu} = 0$

It is shown in this spinor representation, that setting either factor in equation (N7) to zero is an equivalent requirement, thus equation

(N7.1) $(\Sigma \gamma_\mu \, \partial/\partial x_\mu + m_0 c/\hbar)\Psi = 0$

is Dirac's famous equation without any electromagnetic terms.

Note that this relativistic equation (modified to include electromagnetic terms) was used to describe electrons and positrons whose rest mass is non-zero.

Moreover, the magnetic moment of the electron is seen to be a relativistic effect.

Particles With Zero Rest Mass Elude Dirac Equation

Again, though, for systems which have zero rest mass, such as photons, gravitons or possibly neutrinos, the second term of equation

(N7.1) $(\Sigma \gamma_\mu \, \partial/\partial x_\mu + m_0 c/\hbar)\Psi = 0$

vanishes ($m_0 = 0$) and the equation cannot easily be modified to include a potential energy term since the only energy term ($m_0 c/\hbar = E_0/c\hbar = 0$) has vanished. This is not too surprising since Dirac started with the Klein-Gordon equation.

Let us now take a look at Schrödinger's non-relativistic wave equation.

Chapter 3
The Schrödinger Equation

Schrödinger's time independent equation

(N8) $-(\hbar^2/(2m_0))\nabla^2\Psi(r,t) + V\Psi(r,t) = E_T\Psi(r,t)$

and Schrödinger's time dependent equation

(N9) $-(\hbar^2/(2m_0))\nabla^2\Psi(r,t) = -\hbar/i(\partial/(\partial t))\Psi(r,t)$

were both developed for systems moving much less than the speed of light in vacuum. Thus, Newtonian dynamics are clearly applicable. Equation (N8) can readily be used to describe stationary states for bound systems which move under the influence of force fields described by a potential energy function V. Note that E_T in equation (N8) is

(N10) $E_T = p^2/(2m_0) + V = E_K + V$

and defined to be the sum of kinetic energy, $E_K = p^2/(2m_0)$ and potential energy, V.

Free Particle Schrödinger Wave Function

A free particle (V = 0) wave function that satisfies Schrödinger's equation is

(N11) $\Psi(r,t) = \Psi_0 e^{(rp - E_K t)i/\hbar}$

where E_K is the kinetic energy given by

(N12) $E_K = p^2/(2m_0) = (½)m_0 v^2$

v is the system's velocity and $p = m_0 v$ is its momentum.

Schrödinger Equation   15

Note that when the particle is at rest ($v = 0$), both the momentum, ($p = m_0 v$) and the kinetic energy, $(½)m_0 v^2 = E_K$ are zero, so that the Schrödinger free particle wave function in equation (N11) correctly collapses to the initial wave amplitude Ψ_0 or

(N11.01) $\Psi(r,t) = \Psi_0 e^{(0-0)} = \Psi_0$

As has been demonstrated, this is in sharp contrast to either the corresponding free particle Klein–Gordon or a Dirac wave function of equation

(N4) $\Psi(r,t) = \Psi_0 e^{(rp - Et)i/\hbar}$

Recalling that for zero momentum and rest energy E_0 has been shown to be

(N4.3) $\Psi(0,t) = \Psi_0 e^{-(E_0 t)i/\hbar}$

which does not collapse for stationary free particles having non-zero rest mass.

Origin of Schrödinger's Equations

Let us further investigate the origin of Schrödinger's famous equations. We will begin with the free particle wave function described by

(N11) $\Psi(r,t) = \Psi_0 e^{(rp - E_K t)i/\hbar}$

which may be simplified by considering the free particle moving down the x axis. Equation (N11) then becomes

(N11.1) $\Psi(x,t) = \Psi_0 e^{(xp_x - E_K t)i/\hbar}$

Taking a partial derivative of the wave function of (N11) with respect to time yields

(N11.2) $\partial \Psi(r,t)/\partial t = (-i/\hbar) E_K \Psi(r,t)$

which can be rewritten as

(N11.3) $-(\hbar/i) \partial \Psi(r,t)/\partial t = E_K \Psi(r,t)$

Taking two partial derivatives with respect to x of equation (N11.1) yields

(N11.4) $\partial^2 \Psi(x,t)/\partial x^2 = -(p_x/\hbar)^2 \Psi(x,t)$

and because of

(N12) $E_K = p_x^2/(2m_0) = (½)m_0 v_x^2$

can be rewritten as

(N11.5) $(-\hbar^2/(2m_0))\partial^2 \Psi(x,t)/\partial x^2 = E_K \Psi(x,t)$

Time Independent Equation

Converting (N11.5) back into three dimensions, yields,

(N11.6) $-(\hbar^2/(2m_0))\nabla^2 \Psi(r,t) = E_K \Psi(r,t)$

This can be further generalized by adding a potential energy, $V\Psi(r,t)$ term to both sides yielding

$$\text{(N11.7)} \; -(\hbar^2/(2m_0))\nabla^2\Psi(r,t) + V\Psi(r,t) = E_T\Psi(r,t)$$

where the definition of total energy, via (N10) $E_T = E_K + V = p^2/(2m_0) + V$ has been utilized. This exactly reproduces Schrödinger's time independent (no time derivatives) equation given by.

$$\text{(N8)} \; -(\hbar^2/(2m_0))\nabla^2\Psi(r,t) + V\Psi(r,t) = E_T\Psi(r,t)$$

Note that that time dependence of the wave function can be completely removed. If one plugs the wave function of equation

$$\text{(N11)} \; \Psi(r,t) = \Psi_0 e^{(rp - E_K t)i/\hbar}$$
$$= \Psi_0 e^{(rp)i/\hbar} e^{(-E_K t)i/\hbar}$$

back into (N11.7), the time dependence, $e^{(-E_K t)i/\hbar}$ divides out and equation (N11.7) may be rewritten as

(N11.9) $-(\hbar^2/(2m_0))\nabla^2\Psi(r) + V\Psi(r) = E_T\Psi(r)$

of which all reference to time has been eliminated. The solutions of equation (N11.9) represents the so called time independent stationary states of the particle. Even though equation (N11.9) is the formal time independent equation, equation

(N8) $-(\hbar^2/(2m_0))\nabla^2\Psi(r,t) + V\Psi(r,t) = E_T\Psi(r,t)$

will still be referred to as Schrödinger's time independent equation since it is known how the time dependence may be completely eliminated.

Time Dependent Equation

Utilizing equation

(N11.6) $-(\hbar^2/(2m_0))\nabla^2\Psi(r,t) = E_K\Psi(r,t)$

with equation

20 An Alternative Relativistic Wave Equation

(N11.3) $-(\hbar/i)\, \partial\Psi(r,t)/\partial t = E_K \Psi(r,t)$

produces

(N11.8) $-(\hbar/i)\partial\Psi(r,t)/\partial t = -(\hbar^2/(2m_0))\nabla^2\Psi(r,t)$

which can be rewritten as

(N11.81) $-(\hbar^2/(2m_0))\nabla^2\Psi(r,t) + (\hbar/i)\partial\Psi(x,t)/\partial t = 0$

exactly reproducing Edwin Schrödinger's time dependent equation

(N9) $-(\hbar^2/(2m_0))\nabla^2\Psi(r,t) = -\hbar/i(\partial/(\partial t)\Psi(r,t)$

proposed in 1926.

Operator Relationships

If one defines the Hamiltonian operator, $\hat{H}$ as the sum of a kinetic energy operator, $\check{T}$ and the potential energy, V as $\hat{H} = \check{T} + V$ then the time independent equation

(N8) $-(\hbar^2/(2m_0))\nabla^2\Psi(r,t) + V\Psi(r,t) = E_T\Psi(r,t)$

can be rewritten in operator notation as

(N8.1) $\hat{H}\Psi(r,t) = (\check{T} + V)\Psi(r,t) = E_T\Psi(r,t)$

where

(N8.2) $\check{T} = -(\hbar^2/(2m_0))\nabla^2$

Utilizing the time dependent equation

(N11.9) $-(\hbar^2/(2m_0))\nabla^2\Psi(r,t) + (\hbar/i)\partial\Psi(x,t)/\partial t = 0$

means that the kinetic energy operator can be rewritten as

(N11.9.1) $\check{T} = -(\hbar/i)\partial/\partial t$

which means that equation

(N11.3) $-(\hbar/i)\, \partial\Psi(r,t)/\partial t = E_K \Psi(r,t)$

may be rewritten as

(N11.3.1) $\check{T}\Psi(r,t) = E_K\Psi(r,t)$

The non-relativistic Schrödinger equations are seen to be self consistent. The time independent equation is perfect for describing the non-relativistic electrons within the atom.

Let us now present some relativistic preliminaries to help develop the tools that will be used to derive a more general relativistic wave equation.

Chapter 4
Relativistic Preliminaries

Assume a relativistic energy system having a four vector position as

(P1) s = (x, y, z, ict)

with initial four vector position as

(P2) s₀ = (0, 0, 0, 0)

and having a four vector momentum

(P3) p = (p_x, p_y, p_z, iE/c)

with initial four vector momentum
(P4) p₀ = (0, 0, 0, iE₀/c)

where (x, y, z) is the three dimensional position and t is when the system was at (x, y, z). c is the velocity of light in vacuum and E is its total energy given by

(N4.1) $E = mc^2$

with rest mass energy again as

(N5) $E_0 = m_0 c^2$

Relativistic Action Tool

Form the vector scalar product of four vector position and four vector momentum to get an expression for the action (defined to be either the product of momentum and position or energy and time). This action is known to represent the oscillatory part of the wave function. Performing the operation yields

(P5) $(s - s_0) \bullet (p - p_0) = x p_x + y p_y + z p_z - (E - E_0) t.$

Now choose the system moving down the x axis so that the right hand side of equation (P5) becomes

(P6) $(s - s_0) \cdot (p - p_0) = xp - (E - E_0)t$

where $p_x = p$, $p_y = 0$, $p_z = 0$.

The tool of equation (P6) represents the oscillatory part of a relativistic particle's wave function.

Let us now see how equation

(P6) $(s - s_0) \cdot (p - p_0) = xp - (E - E_0)t$

is connected to a relativistic free particle.

Chapter 5
Free Particle Derivation

Consider the corresponding wave described by equation

(P6) $(s - s_0) \cdot (p - p_0) = xp - (E - E_0)t$

moving down the x axis and described by a time dependent relativistic wave function

(D1) $\Psi(x,t) = \Psi_0 e^{(s - s_0) \cdot (p - p_0)i/\hbar}$

$\qquad = \Psi_0 e^{(xp - (E - E_0)t)i/\hbar}$

where $\Psi_0 = \Psi(0, 0)$ is the initial wave amplitude.

Note that in equation (D1) the kinetic energy is represented by $E - E_0 = (m - m_0)c^2$ rather than

$p^2/(2m_0) = (½)m_0v^2$. Note also, that the wave function oscillates only when the system moves. If the system is at rest, then the wave function consistently collapses to the initial wave amplitude Ψ_0.

Taking a partial derivative with respect to x of the wave function in equation (D1), we obtain

(D1.1) $\partial\Psi(x,t)/\partial x = (ip/\hbar)\Psi(x,t)$

Classic Momentum Operator

This can be generalized to three dimensions by replacing the spatial derivative with the ∇ operator of equation (N2) $\nabla = (\partial/\partial x, \partial/\partial y, \partial/\partial z)$ and replacing (x) with (x,y,z) = (r) to yield

(D1.2) $\nabla\Psi(r,t) = (ip/\hbar)\Psi(r,t)$

which preserves the classic momentum operator relationship namely

(D1.3) $(\hbar/i)\nabla\Psi(r,t) = p\Psi(r,t)$

Free Particle Time Independent Wave Equation

Taking a second partial of x derivative of equation (D1.2) produces

(D2) $\partial^2\Psi(x,t)/\partial x^2 = -(p^2/\hbar^2)\Psi(x,t)$

However, a known relativistic relationship holds between momentum p, energy E and rest energy E_0, namely that (D3) $p^2c^2 = E^2 - E_0^2$ therefore, equation (D2) becomes

(D2.1) $\partial^2\Psi(x,t)/\partial x^2 = -((E^2-E_0^2)/(c^2\hbar^2))\Psi(x,t)$
$= -((E-E_0)(E+E_0)/(c^2\hbar^2))\Psi(x,t)$

and can be rewritten as

(D2.2) $(-\hbar^2/(m + m_0))\partial^2\Psi(x,t)/\partial x^2 = (E-E_0)\Psi(x,t)$

which may be generalized, by replacing the spatial partial derivatives with the normal differential vector operator ∇ and replacing (x) with (x,y,z) = (r) to get

(D2.3) $(-\hbar^2/(m + m_0))\nabla^2\Psi(r,t) = (E-E_0)\Psi(r,t)$

Relativistic Operators For Free Particles

Note that in terms of operators, equation (D2.3) is expressed as

(D3.1) $\check{T}\Psi(r,t) = (E-E_0)\Psi(r,t)$

where $\check{T}$ is the kinetic energy operator defined by

(D3.2) $\check{T} = (-\hbar^2/(m + m_0))\nabla^2$

Invariant Forms

Equation

$$(D2.3) \; (-\hbar^2/(m + m_0))\nabla^2\Psi(r,t) = (E-E_0)\Psi(r,t)$$

can be rewritten in invariant form by subtracting $E\Psi(r,t)$ from both sides and then multiplying through by a (−1) to get

$$(D3.3) \; (\hbar^2/(m + m_0))\nabla^2\Psi(r,t) + E\Psi(r,t) = E_0\Psi(r,t)$$

which will work in any Lorentz coordinate system.

The Hamiltonian

Another way to look at equation (D2.3) is to rewrite it as equation

$$(D3.4) \; (-\hbar^2/(m + m_0))\nabla^2\Psi(r,t) + E_0\Psi(r,t) = E\Psi(r,t)$$

In this form, it is apparent that the Hamiltonian operator for a free relativistic particle is

(D4) $\hat{H} = (-\hbar^2/(m + m_0))\nabla^2 + E_0$

Note that it consistently follows the form $\hat{H} = \check{T} + V$ where

(D4.1) $\check{T} = (-\hbar^2/(m + m_0))\nabla^2$

and

(D4.2) $V = E_0$

Let us now see how to further generalize equation

(D2.3) $(-\hbar^2/(m + m_0))\nabla^2\Psi(r,t) = (E-E_0)\Psi(r,t)$

Chapter 6
Time Independent Equation

Further generalization may be achieved by considering the particle in a field of force described by a potential energy, V. Defining the total energy as the sum of kinetic energy, T and potential energy as

(D4.3) $E_T = T + V$

where the relativistic kinetic energy, T is

(D4.4) $T = E - E_0 = (m - m_0)c^2$

equation

(D2.3) $(-\hbar^2/(m + m_0))\nabla^2\Psi(r,t) = (E-E_0)\Psi(r,t)$

becomes

(D5) $(-\hbar^2/(m + m_0))\nabla^2\Psi(r,t) + V\Psi(r,t) = E_T\Psi(r,t)$

Equation (D5) can be used to describe stationary states of systems which have zero rest mass, such as photons, gravitons or possibly neutrinos which may move under the influence of forces described by a potential energy function V.

Recall that even though photons have zero rest mass, they nevertheless, have mass when they are moving, given by $m = \hbar\omega/c^2$ where ω is the angular frequency of the photon.

Operator Representation

Equation

(D5) $(-\hbar^2/(m + m_0))\nabla^2\Psi(r,t) + V\Psi(r,t) = E_T\Psi(r,t)$

can be rewritten in operator form as

(D5.1) $\hat{H}\Psi(r,t) = E_T\Psi(r,t)$

where the Hamiltonian operator is

(D5.2) $\hat{H} = (-\hbar^2/(m + m_0))\nabla^2 + V$

which means that the kinetic energy operator is

(D5.3) $\check{T} = (-\hbar^2/(m + m_0))\nabla^2$

which is the same as for a free relativistic particle.

We will now proceed to obtain a time dependent relativistic equation which will complement the derived time independent equation

(D5) $(-\hbar^2/(m + m_0))\nabla^2\Psi(r,t) + V\Psi(r,t) = E_T\Psi(r,t)$

Chapter 7
Towards The Time Dependent Equation

Taking a partial derivative with respect to t of the wave function in equation

(D1) $\Psi(x,t) = \Psi_0 e^{(s-s_0)\cdot(p-p_0)i/\hbar}$

$\qquad = \Psi_0 e^{(xp-(E-E_0)t)i/\hbar}$

we obtain

(D6) $-(\hbar/i)\partial\Psi(x,t)/\partial t = (E-E_0)\Psi(x,t)$

which in three dimensions becomes;

An Alternative Relativistic Wave Equation

$$\text{(D6.1)} \ -(\hbar/i)\partial\Psi(r,t)/\partial t = (E-E_0)\Psi(r,t)$$

and when influenced by a potential energy function V, can be further generalized by combining equations

$$\text{(D4.3)} \ E_T = T + V \ \text{and}$$

$$\text{(D4.4)} \ T = E - E_0 \ \text{resulting in}$$

$$\text{(D4.5)} \ E - E_0 + V = E_T$$

Plugging (D4.5) into (D6.1) results in

$$\text{(D6.2)} \ -(\hbar/i)\partial\Psi(r,t)/\partial t + V\Psi(r,t) = E_T\Psi(r,t)$$

Taking two partial derivatives of equation

$$\text{(D1)} \ \Psi(x,t) = \Psi_0 e^{(s-s_0)\bullet(p-p_0)i/\hbar}$$
$$= \Psi_0 e^{(xp-(E-E_0)t)i/\hbar}$$

with respect to time, t and dividing by c^2, in three dimensions yields

(D7) $\partial^2\Psi(r,t)/(c^2\partial t^2) = -((E-E_0)^2/(c^2\hbar^2))\Psi(r,t)$
$= -((E^2-2EE_0+E_0^2)/(c^2\hbar^2))\Psi(r,t)$

Note that equation

(D2.1) $\partial^2\Psi(x,t)/\partial x^2 = -((E^2-E_0^2)/(c^2\hbar^2))\Psi(x,t)$

can be rewritten and generalized to three dimensions becoming

(D2.3) $\nabla^2\Psi(r,t) = -((E^2-E_0^2)/(c^2\hbar^2))\Psi(r,t)$

Subtracting equation (D7) from equation (D2.3) we obtain

(D8) $\nabla^2\Psi(r,t) - \partial^2\Psi(r,t)/(c^2\partial t^2) =$
$-(2E_0(E-E_0)/(c^2\hbar^2))\Psi(r,t)$

but since $E_0 = m_0c^2$ we get

(D8.1) $(-\hbar^2/(2m_0))(\nabla^2 - \partial^2/(c^2 \partial t^2))\Psi(r,t) = (E-E_0)\Psi(r,t)$

and by equation

(N3.1) $\square \bullet \square = \square^2 = \nabla^2 - \partial^2/(c^2 \partial t^2)$

becomes

(D8.2) $(-\hbar^2/(2m_0))\square^2 \Psi(r,t) = (E-E_0)\Psi(r,t)$

Kinetic Energy Operator

Note that equation (D8.2) can be rewritten as

(D8.3) $\check{T}\Psi(r,t) = (E-E_0)\Psi(r,t)$

where the kinetic energy operator is

(D8.4) $\check{T} = (-\hbar^2/(2m_0))\square^2$

Let us now generalize equation

(D8.2) $(-\hbar^2/(2m_0))\Box^2\Psi(r,t) = (E-E_0)\Psi(r,t)$

to arrive at a time dependent relativistic wave equation. This resulting equation may then be compared to the Klein-Gordon equation.

Chapter 8
Time Dependent Equation

Equation

(D8.2) $(-\hbar^2/(2m_0))\Box^2\Psi(r,t) = (E-E_0)\Psi(r,t)$

can be further generalized via equation

(D4.5) $E - E_0 + V = E_T$

if influenced by a potential energy function V, as

(D9) $(-\hbar^2/(2m_0))\Box^2\Psi(r,t) + V\Psi(r,t) = E_T\Psi(r,t)$

This is the relativistic time dependent equation.

Rewriting equation (D9) eliminating the potential energy term V by equation

(D4.5) $E - E_0 + V = E_T$

yields

(D9.1) $(\hbar^2/(2m_0))\Box^2\Psi(r,t) + E\Psi(r,t) = E_0\Psi(r,t)$

Comparing this with the corresponding Klein-Gordon equation rewritten as

(N1.3.1) $(\hbar^2/m_0)\Box^2\Psi(r,t) = E_0\Psi(r,t)$

both written in invariant form, reveals the missing piece of the Klein-Gordon equation. It is missing the total dynamic energy ($E = mc^2$) term, $E\Psi(r,t)$.

Generalized Wave Equations

Thus, this derived equation

(D9) $(-\hbar^2/(2m_0))\Box^2\Psi(r,t) + V\Psi(r,t) = E_T\Psi(r,t)$

is the time dependent complement to equation

(D5) $(-\hbar^2/(m + m_0))\nabla^2\Psi(r,t) + V\Psi(r,t) = E_T\Psi(r,t)$

Together equations (D5) and (D9) represent the desired new alternative Lorentz invariant, relativistic wave equations.

The other first order time dependent equation

(D6.2) $-(\hbar/i)\partial\Psi(r,t)/\partial t + V\Psi(r,t) = E_T\Psi(r,t)$

contains the term for the evolution of the relativistic wave function and is the basis for linear causality.

Three Forms of the Relativistic Hamiltonian

The Relativistic Hamiltonian (total energy) Operator may be defined as a sum of two terms as

(D10) $\hat{H} = \check{T} + V$

where $\check{T}$ is defined to be the kinetic energy operator and V is the potential energy.

In the time independent equation

(D5) $(-\hbar^2/(m + m_0))\nabla^2\Psi(r,t) + V\Psi(r,t) = E_T\Psi(r,t)$

recall equation

(D5.1) $\hat{H}\Psi(r,t) = E_T\Psi(r,t)$

and with the Hamiltonian operator equation

(D10) $\hat{H} = \check{T} + V$

reproduces equation

(D5.2) $\hat{H} = (-\hbar^2/(m + m_0))\nabla^2 + V$

In equation

An Alternative Relativistic Wave Equation

(D6.2) $-(\hbar/i)\partial\Psi(r,t)/\partial t + V\Psi(r,t) = E_T\Psi(r,t)$

the kinetic energy operator is

(D6.3) $\check{T} = -(\hbar/i)\partial/\partial t$

and so the Hamiltonian (equation (D10)) becomes

(D6.4) $\hat{H} = -(\hbar/i)\partial/\partial t + V$

For the time dependent equation

(D9) $(-\hbar^2/(2m_0))\Box^2\Psi(r,t) + V\Psi(r,t) = E_T\Psi(r,t)$

the kinetic energy operator is

(D9.2) $\check{T} = (-\hbar^2/(2m_0))\Box^2$

and the Hamiltonian operator becomes

(D9.3) $\hat{H} = (-\hbar^2/(2m_0))\Box^2 + V$

Summary of Hamiltonians

Thus, for the time independent equation, the Hamiltonian operator is

(D5.2) $\hat{H} = (-\hbar^2/(m + m_0))\nabla^2 + V$

For the time dependent equations, the Hamiltonians are

(D6.4) $\hat{H} = -(\hbar/i)\partial/\partial t + V$

and

(D9.3) $\hat{H} = (-\hbar^2/(2m_0))\Box^2 + V$

Let us now examine how a relativistic wave function evolves in time.

Chapter 9
Evolution of the Wave Function

Equating the time independent equation

(D5) $(-\hbar^2/(m + m_0))\nabla^2\Psi(r,t) + V\Psi(r,t) = E_T\Psi(r,t)$

with the time dependent equation

(D6.2) $-(\hbar/i)\partial\Psi(r,t)/\partial t + V\Psi(r,t) = E_T\Psi(r,t)$

yields

(D10.1) $-(\hbar/i)\partial\Psi(r,t)/\partial t = -(\hbar^2/(m + m_0))\nabla^2\Psi(r,t)$

Note that this equation has a term, $\partial\Psi(r,t)/\partial t$ which controls the time evolution of the wave function. In fact, solving for this in equation (D10.1) produces

(D10.2) $\partial\Psi(r,t)/\partial t = (i\hbar/(m + m_0))\nabla^2\Psi(r,t)$

Equating equation

(D9) $(-\hbar^2/(2m_0))\square^2\Psi(r,t) + V\Psi(r,t) = E_T\Psi(r,t)$

with equation

(D6.2) $-(\hbar/i)\partial\Psi(r,t)/\partial t + V\Psi(r,t) = E_T\Psi(r,t)$

yields

(D10.3) $-(\hbar/i)\partial\Psi(r,t)/\partial t = (-\hbar^2/(2m_0))\square^2\Psi(r,t)$

Note that this equation also has a term, $\partial\Psi(r,t)/\partial t$ which controls the time evolution of the wave function. In fact, solving for this in equation (D10.3) produces

(D10.4) $\partial \Psi(r,t)/\partial t = (i\hbar/(2m_0))\square^2\Psi(r,t)$

which is very similar to

(D10.2) $\partial \Psi(r,t)/\partial t = (i\hbar/(m+m_0))\nabla^2\Psi(r,t)$

Wave Function Analysis of a Particle at Rest

Equations (D10.4) and (D10.2) yield two prescriptions of how the wave function changes in time. By inspection, if the particle is at rest, then its mass is equal to it rest mass.

In fact putting $m = m_0$ in equation (D10.2), and subtracting equation D(10.4) yields

(D10.5) $\square^2\Psi(r,t) = \nabla^2\Psi(r,t)$

Equation (D10.5) is only possible if

(D10.6) $\partial^2 \Psi(r,t)/\partial t^2 = 0$

which is true if either the wave function does not depend on time, t or that

(D10.7) $\partial \Psi(r,t)/\partial t = \text{constant} = K_1$

Integrating, we get that

(D10.8) $\Psi(r,t) = K_1 t + \Psi(0,0)$

Clearly, this wave function increases without bound linearly with respect to time. It cannot be normalized. Thus, since this possibility is eliminated, it follows that the wave function, $\Psi(r,t)$ is independent of time and can be written as

(D10.9) $\Psi(r,t) = \Psi(r)$

so as to ensure the result of equation

50 An Alternative Relativistic Wave Equation

(D10.6) $\partial^2 \Psi(r,t)/\partial t^2 = 0$

Thus, we have established the quantum mechanical meaning of equation

(D10.6) $\partial^2 \Psi(r,t)/\partial t^2 = 0$

in that it implies the particle is at rest and the wave function is independent of time. With this result, it is seen that the time independent equation

(D5) $(-\hbar^2/(m + m_0))\nabla^2 \Psi(r,t) + V\Psi(r,t) = E_T\Psi(r,t)$

is equivalent to the time dependent equation

(D9) $(-\hbar^2/(2m_0))\Box^2 \Psi(r,t) + V\Psi(r,t) = E_T\Psi(r,t)$

if the particle is at rest with $m = m_0$.

Wave Function Analysis of a Particle in Motion

For moving particles, then equations

(D10.2) $\partial\Psi(r,t)/\partial t = (i\hbar/(m + m_0))\nabla^2\Psi(r,t)$

and

(D10.4) $\partial\Psi(r,t)/\partial t = (i\hbar/(2m_0))\square^2\Psi(r,t)$

yield two ways to compute the time evolution of the wave function, $\Psi(r,t)$.

Equation (D10.2) may be used when the velocity of the particle is known and the time dependency of the wave function is unknown.

Equation (D10.4) is used when the velocity of the particle is not known and the time dependency of the wave function is known.

Of course equation

52 An Alternative Relativistic Wave Equation

(D6.2) $-(\hbar/i)\partial\Psi(r,t)/\partial t + V\Psi(r,t) = E_T\Psi(r,t)$

can be solved for $\partial\Psi(r,t)/\partial t$ which yields

(D6.2.1) $\partial\Psi(r,t)/\partial t = -(i/\hbar)(E_T - V)\Psi(r,t)$

which also may be used to compute the time evolution of the wave function when the total energy, E_T and the potential energy, V are both known.

Note that equation (D6.2.1) can also be rewritten as

(D6.2.1.1) $\partial\Psi(r,t)/\partial t = -(i/\hbar)T\Psi(r,t)$ or

(D6.2.1.2) $\partial\Psi(r,t)/\partial t = -(i/\hbar)(E - E_0)\Psi(r,t)$ since

(D4.5) $E - E_0 + V = E_T$ and

(D4.4) $T = E - E_0 = (m - m_0)c^2$

Chapter 10
Conservation of Probability

Probability density, ρ_P by the Born interpretation, is given by

(D11) $\rho_P = \Psi^*\Psi$

where the notation has been simplified to mean

(D11.01) $\Psi = \Psi(r,t)$

and Ψ^* is the complex conjugate of Ψ.

This means that the infinitesimal probability, dP that a particle represented by a wave function, Ψ is located in an infinitesimal volume, d^3r is given by

(D11.1) $dP = \rho_P d^3r$

Conservation of Relativistic Probability Current

The rate of change of the probability density is given by

(D11.2) $\partial \rho_P/\partial t = \Psi \partial \Psi^*/\partial t + \Psi^* \partial \Psi/\partial t$

Utilizing equation

((D10.2) $\partial \Psi(r,t)/\partial t = (i\hbar/(m + m_0))\nabla^2 \Psi(r,t)$

whose complex conjugate is

(D10.2.1) $\partial \Psi^*(r,t)/\partial t = -(i\hbar/(m + m_0))\nabla^2 \Psi^*(r,t)$

Plugging both equations (D10.2) and (D10.2.1) into equation (D11.2) yields

Conservation of Probability 55

(D11.2.1) $\partial\rho_P/\partial t = -\Psi(r,t)(i\hbar/(m+m_0))\nabla^2\Psi^*(r,t)$
$+ \Psi(r,t)^*(i\hbar/(m+m_0))\nabla^2\Psi(r,t)$

which can be rewritten as

(D11.3) $\partial\rho_P/\partial t = -[i\hbar/(m+m_0)]\nabla\bullet\{\Psi(r,t)\nabla\Psi^*(r,t)$
$- \Psi(r,t)^*\nabla\Psi(r,t)\}$

where the relativistic probability current vector, $\check{I}_P$ is defined as

(D12) $\check{I}_P = -[i\hbar/(m+m_0)]\{\Psi\nabla\Psi^*(r,t) - \Psi^*\nabla\Psi(r,t)\}$

which makes equation

(D11.3) $\partial\rho_P/\partial t = -[i\hbar/(m+m_0)]\nabla\bullet\{\Psi\nabla\Psi^*(r,t) -$
$\Psi^*\nabla\Psi(r,t)\}$

simplify to

(D11.4) $\partial \rho_P/\partial t = -\nabla \bullet \breve{I}_P$

which can finally be rewritten as

(D11.5) $\partial \rho_P/\partial t + \nabla \bullet \breve{I}_P = 0$

which means that the relativistic probability current is conserved.

Rate of Change of Expectation Values

The expectation value of a dynamical variable Y is defined to be

(D14) $\langle Y \rangle = \iiint \Psi^* \hat{Y} \Psi d^3r$

where $\Psi = \Psi(r,t)$, Ψ^* is the complex conjugate of Ψ, $d^3r = dxdydz$ and $\hat{Y}$ is the corresponding operator for the dynamical variable Y. Taking a time derivative of (D14) results in

(D14.1) $d\langle Y\rangle/dt = \iiint(\partial\Psi^*/\partial t)\hat{Y}\Psi d^3r +$
$\iiint\Psi^*(\partial\hat{Y}/\partial t)\Psi d^3r + \iiint\Psi^*\hat{Y}(\partial\Psi/\partial t)d^3r$

Consider equation

(D6.2.1) $\partial\Psi/\partial t = -(i/\hbar)(E_T - V)\Psi$

and also its complex conjugate

(D6.2.2) $\partial\Psi^*/\partial t = (i/\hbar)(E_T - V)\Psi^*$

Substituting both equations (D6.2.1) and (D6.2.2) into (D14.1) results in

(D14.2) $d\langle Y\rangle/dt = \langle\partial Y/\partial t\rangle +$
$(i/\hbar)\iiint(E_T - V)\Psi^*\hat{Y}\Psi d^3r -$
$(i/\hbar)\iiint\Psi^*\hat{Y}(E_T - V)\Psi d^3r$

which reduces to

58 An Alternative Relativistic Wave Equation

(D14.3) $d\langle Y\rangle/dt = \langle\partial Y/\partial t\rangle +$

$$(i/\hbar)\langle(E_T - V)\hat{Y} - \hat{Y}(E_T - V)\rangle$$

and because of equation (D4.5) $E - E_0 + V = E_T$ and recalling the definition of relativistic kinetic energy, T as

(D4.4) $T = E - E_0 = (m - m_0)c^2$

and with

(D10) $\hat{H} = \check{T} + V$

equation (D14.3) can be rewritten as

(D14.4) $d\langle Y\rangle/dt = \langle\partial Y/\partial t\rangle + (i/\hbar)\langle\hat{Y}\check{T} - \check{T}\hat{Y}\rangle$

Let us now see how this tool will help establish some important results associated with the derived wave equations.

Chapter 11
Quantum Mechanical Implications

Statistical Conservation of Energy

If the Hamiltonian operator in equation

(D6.4) $\hat{H} = -(\hbar/i)\partial/\partial t + V$

is substituted into equation

(D14.4) $d\lessgtr Y\gtrless/dt = \lessgtr \partial Y/\partial t \gtrless + (i/\hbar)\lessgtr \hat{Y}\check{T} - \check{T}\hat{Y}\gtrless$

for the operator $\hat{Y}$, the result yields the statistical conservation of total energy

(D17) $d\langle E_T\rangle/dt = \langle\partial E_T/\partial t\rangle + (i/\hbar)\langle 0\rangle = 0$

since E_T is independent of time and since $\hat{H}$ commutes with the potential energy, V provided V is not a function of time.

Persistence of Normalization

Moreover, if the unity operator (= 1) is substituted into equation

(D14.4) $d\langle Y\rangle/dt = \langle\partial Y/\partial t\rangle + (i/\hbar)\langle\hat{Y}\check{T} - \check{T}\hat{Y}\rangle$

for the operator $\hat{Y}$, then by equation

(D14) $\langle Y\rangle = \iiint \Psi^*\hat{Y}\Psi d^3r$

equation (D14.4) becomes

(D15) $d(\iiint \Psi^*\Psi d^3r)/dt = 0$

which establishes that the normalization integral, $\iiint \Psi^*\Psi d^3r$ is independent of time.

Equation

(D1.3) $(\hbar/i)\nabla\Psi(r,t) = p\Psi(r,t)$

demonstrates that the classic momentum operator, $(\hbar/i)\nabla$ is preserved in the relativistic case.

Thus, the newly derived Lorentz Invariant Relativistic Equations are compatible with basic quantum mechanics. Conservation of energy, conservation of probability and the persistence of normalization have all been consistently verified. The form of the momentum operator is also preserved.

Chapter 12
Summary and Conclusions

Time Independent Equation

Equation

(D5) $(-\hbar^2/(m + m_0))\nabla^2\Psi(r,t) + V\Psi(r,t) = E_T\Psi(r,t)$

is a Lorentz invariant relativistic time independent wave equation that may be used to describe systems that have zero rest mass and move under the influence of a potential V. Note that in equation (D5), m represents the system's mass in motion, while m_0 represents the same system's mass at rest. In other words, $m = m_0(1-(v/c)^2)^{-1/2}$ where v is the system's velocity.

Reduction to Schrödinger's Time Independent Equation

Equation (D5) readily and cleanly reduces to Schrödinger's time independent equation

(N8) $-(\hbar^2/(2m_0))\nabla^2\Psi(r,t) + V\Psi(r,t) = E_T\Psi(r,t)$

for if a system's velocity is small compared to the velocity of light c, means that $m \approx m_0$ and $m + m_0 \approx 2m_0$. Plugging this into equation (D5) results in equation (N8).

Time Dependent Equations

Equation

(D9) $(-\hbar^2/(2m_0))\Box^2\Psi(r,t) + V\Psi(r,t) = E_T\Psi(r,t)$

is a Lorentz invariant relativistic time dependent wave equation that is an alternate to either the Klein-Gordon wave equation given by

64   An Alternative Relativistic Wave Equation

(N1.2) $(-\hbar^2/m_0)\Box^2\Psi(r,t) = E_0\Psi(r,t)$

or the Dirac wave equation

(N7.1) $(\Sigma\gamma_\mu \,\partial/\partial x_\mu + m_0c/\hbar)\Psi = 0$

where $(x_1, x_2, x_3, x_4) = (x, y, z, ict)$

and $\gamma_\mu\gamma_\nu + \gamma_\nu\gamma_\mu = 2\delta_{\mu\nu}$

Both the Klein-Gordon as well as the Dirac equation are time dependent. Therefore, in order to get a time independent form (so that stationary states can be described) requires external assumptions about the periodic behavior of the wave function.

The derivation of Schrödinger's equations from either the Klein-Gordon or Dirac equations require much imagination along with a host of external assumptions.

The other time dependent equation is

(D6.2) $-(\hbar/i)\partial\Psi(r,t)/\partial t + V\Psi(r,t) = E_T\Psi(r,t)$

which contains the time evolution term, $\partial\Psi(r,t)/\partial t$ of the wave function without any second order derivatives. This equation can also be interpreted as the statement for the linear nature of relativistic quantum mechanical causality.

Derived Equations Summary

The three derived equations

(D5) $(-\hbar^2/(m + m_0))\nabla^2\Psi(r,t) + V\Psi(r,t) = E_T\Psi(r,t)$

(D6.2) $-(\hbar/i)\partial\Psi(r,t)/\partial t + V\Psi(r,t) = E_T\Psi(r,t)$ and

(D9) $(-\hbar^2/(2m_0))\Box^2\Psi(r,t) + V\Psi(r,t) = E_T\Psi(r,t)$

all define the three following Hamiltonians.

Hamiltonian Summary

(D5.2) $\hat{H} = (-\hbar^2/(m + m_0))\nabla^2 + V$

(D6.4) $\hat{H} = -(\hbar/i)\partial/\partial t + V$

(D9.3) $\hat{H} = (-\hbar^2/(2m_0))\Box^2 + V$

Kinetic Energy Operator Summary

Each of the above Hamiltonians contain the corresponding kinetic energy operators of equations

(D5.3) $\check{T} = (-\hbar^2/(m + m_0))\nabla^2$

(D6.3) $\check{T} = -(\hbar/i)\partial/\partial t$

(D9.2) $\check{T} = (-\hbar^2/(2m_0))\Box^2$

Conclusions

The derived relativistic equations adhere to the philosophy that if a system is at rest, its wave function collapses.

Analysis of equation

(D14.4) $d\langle Y\rangle/dt = \langle\partial Y/\partial t\rangle + (i/\hbar)\langle \hat{Y}\check{T} - \check{T}\hat{Y}\rangle$

for the rate of change of quantities represented by the operator, $\hat{Y}$ means that operators whose eigenvalues are constants of the motion like energy, momentum, angular momentum and spin must all commute with the kinetic energy operator of the system. In the non-relativistic Schrödinger theory, constants of the motion must commute with the Hamiltonian. In this relativistic theory, if the potential energy, V is not a function of time, then it can be shown that if an operator commutes with the kinetic energy operator, $\check{T}$ then it also commutes with the Hamiltonian.

Equation

(D10.6) $\partial^2\Psi(r,t)/\partial t^2 = 0$

means that either the particle is at rest or that the wave function is independent of time. Time independent wave functions describe the possible stationary energy states of a system. In turn, these states may also be described by the derived time independent wave equation.

Evolution of the Wave Function

The three forms for the evolution of the wave function are equations

(D6.2.1) $\partial\Psi(r,t)/\partial t = -(i/\hbar)(E_T - V)\Psi(r,t)$

Equation (D6.2.1) is used to compute the time evolution of the wave function when the total energy, E_T and the potential energy, V are both known. Note that $E_T - V$ is the kinetic energy, T. This equation is also the basis for the linear nature of relativistic quantum mechanical causality. In other words, knowing a wave function at a certain time t, the wave function can be calculated for any other time assuming the kinetic energy of the system is known.

(D10.2) $\partial\Psi(r,t)/\partial t = (i\hbar/(m + m_0))\nabla^2\Psi(r,t)$

Equation (D10.2) may be used to compute the time evolution of the wave function when the velocity of the particle is known and the time dependency of the wave function is unknown.

(D10.4) $\partial\Psi(r,t)/\partial t = (i\hbar/(2m_0))\Box^2\Psi(r,t)$

Equation (D10.4) is used to compute the time evolution of the wave function when the velocity of the particle is not known and the time dependency of the wave function is known.

GLOSSARY

Action – The product of momentum and position or the product of energy and time. The action part of a wave function define its oscillatory nature.

Commute – Two operators, Ř and Š commute if ŘŠ – ŠŘ = 0.

Dirac wave equation – A time dependent relativistic wave equation which is Lorentz invariant and is capable of describing the electromagnetic properties of an electron and predicts the existence of it anti-particle, the positron.

Eigenvalue – If a quantity (operator) is multiplied (operates on) by the wave function and yields a number times the wave function, then that number is an eigenvalue of that operator.

Free Particle – No external forces act on the particle.

Hamiltonian – An operator which yields the total energy as its eigenvalue.

Klein-Gordon wave equation – A time dependent relativistic wave equation which is Lorentz invariant.

Lorentz Invariant – The form of an equation does not change with respect to a linear coordinate transformation from one four dimensional Euclidean coordinate system moving at a constant velocity with respect to another four dimensional Euclidean coordinate system.

(Planck's Constant)/2π: $h/2\pi = \hbar = 1.0552 \times 10^{-34}$ joule-seconds

Relativistic Kinetic energy – The total dynamic energy, $mc^2 = E$, less the rest mass energy $m_0 c^2 = E_0$

is defined to be the kinetic energy, T. Therefore, $T = E - E_0 = (m - m_0)c^2$.

Schrödinger wave equation – A non-relativistic pair of wave equations. One is time dependent and the other is time independent. These equations are valid for non-relativistic systems (systems which move much less than the speed of light in vacuum). The time independent equation successfully predicts and describes the electron energy levels of atomic nuclei.

Speed of Light: $c = 3.00 \times 10^8$ meters/second

Fundamental Physical Laws

Preliminary Definitions

I. **Bold** mathematical single letters refer to vectors.

II. The symbol, $i = (-1)^{1/2}$ always occurs as the fourth (time component) of all Einsteinian four dimensional vectors.

III. The symbol, c is the speed of light in vacuum.

1. **Position of an energy system:** Referring to Figure 2, a normal Cartesian coordinate system shows the x,y,z position of the system S at time t.

Newtonian: $\mathbf{r}_N = (x,y,z)$

Einsteinian: $\mathbf{r}_E = (x,y,z,ict)$

2. **Velocity of an energy system:** At time t_2, the position of the system was at position 2 (r_2). Initially at time t_1, the position of the system was at position 1 (r_1). The average velocity of the system is the distance traversed by the system in moving from position 1 to position 2 ($r_2 - r_1$) divided by the time it took for the system to move between the two positions ($t_2 - t_1$). The direction of the velocity is from position 1 to position 2. The instantaneous velocity **v** is realized by letting t_2 approach t_1. Mathematically, the instantaneous velocity of a system is a vector quantity.

v = lim as $t_2 \rightarrow t_1$ of $[(\mathbf{r_2} - \mathbf{r_1})/(t_2 - t_1)]$ or

v = d**r**/dt

Newtonian: $\mathbf{v}_N = (v_x, v_y, v_z)$

Einsteinian: $\mathbf{v}_E = (v_x, v_y, v_z, ic)$

3. **Acceleration of an energy system:** At time t_2, the velocity of the system was (v_2). Initially at time

t_1, the velocity of the system was (v_1). The average acceleration of the system is the change in the velocity of the system in going from v_1 to v_2 ($v_2 - v_1$) divided by the time it took for the system to go from v_1 to v_2 ($t_2 - t_1$). The direction of the acceleration is from v_1 to v_2. The instantaneous acceleration **a** is realized by letting t_2 approach t_1. Mathematically, the acceleration of a system is a vector quantity.

$$\mathbf{a} = \lim \text{ as } t_2 \to t_1 \text{ of } [(\mathbf{v_2} - \mathbf{v_1})/(t_2 - t_1)] \quad \text{or}$$

$$\mathbf{a} = d\mathbf{v}/dt$$

Newtonian: $\mathbf{a}_N = (a_x, a_y, a_z)$

Einsteinian: $\mathbf{a}_E = (a_x, a_y, a_z, 0)$

4. **Momentum of an energy system:** The product of the system's mass and its velocity **v** is called its momentum and denoted by **p**. It is a vector quantity having direction **v**.

$\mathbf{p} = m\mathbf{v}$

Newtonian: $\mathbf{p}_N = (p_x, p_y, p_z)$

Einsteinian: $\mathbf{p}_E = (p_x, p_y, p_z, iE/c)$

where E is the total energy = mc^2.

5. **Force on an energy system:** The instantaneous rate of change of a system's momentum with respect to time. Its definition is similar to the definition of velocity. At time t_2 it has momentum 2. At time t_1, it had momentum 1. The average force is the difference in momentum (momentum 2 − momentum 1) divided by the time difference ($t_2 − t_1$). The instantaneous rate is realized when t_2 approaches t_1.

$\mathbf{F} = \lim$ as $t_2 \to t_1$ of $[((m\mathbf{v})_2 − (m\mathbf{v})_1)/(t_2 − t_1)]$ or

$\mathbf{F} = d(m\mathbf{v})/dt$

Newtonian: $\mathbf{F}_N = m_0 d\mathbf{v}/dt = m_0 \mathbf{a}$

where m_0 is the rest mass and **a** is its acceleration.

Einsteinian: $\mathbf{F}_E = d(m\mathbf{v})/dt = md\mathbf{v}/dt + \mathbf{v}dm/dt$ or

$$\mathbf{F}_E = m_0\mathbf{a}(1-(v/c)^2)^{-3/2} = (c^2/v)dm/dt$$

where **v** is m's velocity and **a** is m's acceleration.

6. **Mass density:** Mass m, per unit volume V.

Average mass density = $\rho_{mavg} = m/V$

Instantaneous mass density = $\rho_m = dm/dV$

7. **Pressure on a surface**: The applied force F, per unit surface area, A.

Average pressure = $P_{avg} = F/A$

Instantaneous pressure = $P = dF/dA$

8. **Angular momentum of an energy system:** Let the vector from the origin to the position of the

system be called the position vector (**r**). The angular momentum of the system (**L**) is then the ordinary vector cross product (**x**), of the position vector with the system's momentum vector (m**v**).

L = r x mv

9. **Charge density:** Amount of charge q, per unit volume V.

Average charge density = ρ_{qavg} = q/V

Instantaneous charge density = ρ_q = dq/dV

10. **Electrical current:** The instantaneous change of charge q with respect to time.

i = lim as $t_2 \to t_1$ of $[(q_2 - q_1)/(t_2 - t_1)]$ or

i = dq/dt

11. **Electrical current density:** The electrical current i per unit cross sectional area A of

conductor. The unit vector **u** has a direction of the current i along the conductor perpendicular to the cross sectional area.

$\mathbf{J}_i = \mathbf{u}i/A$ where $i = dq/dt$

In a conductor with conductivity σ_c, the current is in the direction of the electric field **E** and the current density is the product of the conductivity and electric field.

$\mathbf{J}_i = \sigma_c \mathbf{E}$

12. **Mole**: The mass of Avogadro's number of identical molecules or Avogadro's number of identical atoms expressed in grams. One mole of molecules is the molecular weight of the molecule expressed in grams. One mole of atoms is the atomic weight of the atom expressed in grams.

Mechanical Laws

1. **Newton's Laws of Motion:**

1.1 A body will remain at rest or in motion at a constant velocity unless acted on by an unbalanced external force.

1.2 The force on a body is proportional to its acceleration and the constant of proportionality is the rest mass (when the body is at rest), m_0 of the body.

$\mathbf{F} = m_0 \mathbf{a}$

Newton was unaware that mass is a function of its velocity.

1.3 The force of one body on a second body is equal and opposite to the force of the second body on the first body or for every action, there is an equal and opposite reaction.

$\mathbf{F}_{12} = -\mathbf{F}_{21}$

1.4 Newton's Universal Law of Gravitation says that any two energy systems having mass attract each other with a force (**F**) proportional to the product of their masses m_1 and m_2 and inversely proportional to the square of the distance (r) between their mass centers. The force is in a direction between the centers of m_1 and m_2, causing them to attract one another and is denoted by the unit vector $\mathbf{r}_u$. G is the constant of proportionality known as Newton's Gravitational Constant. This force is

$\mathbf{F} = \mathbf{r}_u G m_1 m_2 / r^2$

deriving the gravitational potential energy, V between m_1 and m_2 as

$V = -\mathbf{G} m_1 m_2 / r$

Newtonian: Mass is the cause of the gravitational field.

Einsteinian: Mass energy and momentum warp four dimensional spacetime into a gravitational field.

2. Quantum Mechanical Laws:

2.1 An energy system may be described by a wave function. The total energy operator $\hat{H}$ (known as the Hamiltonian) operating on the wave function (Ψ) yields the total energy eigenvalue (E) of the system represented by the wave function. Energy eigenvalues (E) are the allowable energy states that the system may assume. Similarly, other operators operating on the wave function yield other information (such as the spin, momentum, angular momentum, etc.) about the system.

$\hat{H}\Psi = E\Psi$

2.2 The square of the wave function $\Psi^*\Psi$, multiplied by a infinitesimal volume d^3r is equal to the infinitesimal probability dP, that a system specified by Ψ, is located within that volume.

$dP = \Psi^*\Psi d^3r$

2.3 The probability that an energy system represented by the wave function Ψ, is somewhere in all space is unity, which is the basis for a normalized wave function.

$P = \int dP = \int \Psi^*\Psi d^3r = 1$

3. The Heisenberg Uncertainty Principle:

3.1 In an ideal experiment, the product of the standard deviation in the measurement of a system's momentum, Δp and the standard deviation in the measurement of its position, Δr must be greater than a non-zero constant. This constant is Planck's constant divided by four pi (h/(4π) = ℏ/2) where ℏ is Planck's constant divided by 2π.

$\Delta p \Delta r \geq \hbar/2$

This means that an energy system's position and momentum cannot be known simultaneously.

3.2 Another expression of the Heisenberg uncertainty principle is:

$$\Delta E \Delta t \geq \hbar/2$$

where ΔE is the standard deviation in the measurement of a system's energy and Δt is the standard deviation of the measured times that it had that energy. This means that a system's energy and when it had that energy cannot be known simultaneously.

4. **The energy of a photon (E):** associated with an electromagnetic wave having an angular frequency ω is the product of Planck's rationalized constant ($\hbar$) and its angular frequency. The mass, m_γ is the mass of the photon in flight and c is the speed of light.

$$E = \hbar\omega = m_\gamma c^2$$

5. **De Broglie's relationship:** which expresses that the wavelength of a particle λ is inversely proportional to its momentum. The constant of proportionality is Planck's constant, h.

$$\lambda = h/mv$$

and is sometimes written as

$$\lambdabar = \hbar/mv$$

where $\lambdabar = \lambda/(2\pi)$ and $\hbar = h/(2\pi)$

6. **Einstein's Laws of Special Relativity:** The first four relativistic laws are derived by assuming that the velocity of light c, is independent of the velocity of the source of light as well as the velocity of the observer.

6.1 A system's mass m increases if it is moving with a velocity v compared to the velocity of light c, in vacuum. Initially when the system had a velocity of zero, its rest mass is m_0.

$$m = m_0 \left(1 - (v/c)^2\right)^{-1/2}$$

6.2 A system's length ℓ decreases if it is moving with a velocity v compared to the velocity of light c, in vacuum. Initially when the system had a velocity of zero, its rest length is ℓ_0. ℓ is in the direction of the velocity.

$$\ell = \ell_0 \left(1 - (v/c)^2\right)^{1/2}$$

6.3 A system's clock time, t slows (stretches) if it is moving with a velocity v compared to the velocity of light c, in vacuum. Initially when the system was at rest (had a velocity of zero), it had a clock time of t_0.

$$t = t_0 \left(1 - (v/c)^2\right)^{-1/2}$$

6.4 The total mechanical energy E of a system containing mass is the product of its mass m and the square of the velocity of light c.

$$E = mc^2$$

where $m = m_0(1 - (v/c)^2)^{-1/2}$ is dependent on its velocity v. m_0 (rest mass) is its mass when $v = 0$.

6.5 The relativistic kinetic energy T, of a system in motion is the difference (between its mass in motion less its rest mass) times the velocity of light c, squared.

Einsteinian: $T_E = (m - m_0)c^2$

where $m = m_0(1 - (v/c)^2)^{-1/2}$ is dependent on its velocity, v. For small velocities compared to the velocity of light, the Einsteinian kinetic energy reduces to the Newtonian kinetic energy as a first order approximation. For $v \ll c$, $(m - m_0)c^2 \cong (½)m_0v^2$

Newtonian: $T_N = (½)m_0v^2$

7. Laws of Thermodynamics

7.1 The first law of thermodynamics says that within a closed (isolated) system an amount of heat added to the system dQ results in an increase in its internal energy dU and an amount of work done, dW. Usually, dU results in an increase in internal temperature while dW results in a change in volume dV against a constant pressure p. This also means that energy is conserved for a closed system.

$dQ = dU + dW$ where $dW = pdV$

7.2 The second law of thermodynamics says that a change in the entropy dS of a system undergoing a reversible process is defined to be the amount of heat added dQ divided by its temperature T. If the process is irreversible, then the entropy is always greater than the amount of heat added divided by its temperature.

$dS \geq dQ/T$

where the equality implies reversibility and the greater than symbol (>) implies irreversibility.

7.3 The perfect gas law says that the gas pressure p multiplied by the volume of gas V is proportional to the number of moles n of gas multiplied by the absolute temperature T of the gas. The constant of proportionality R is known as the universal gas constant.

$pV = nRT$

7.4 The fundamental law of heat conduction says that the rate of heat flow dQ/dt across a infinitely thin slab dx of material perpendicular to the surface of the slab is proportional to the surface area A of the slab and the instantaneous absolute temperature change per unit thickness dT/dx of the material. The constant of proportionality K_T is known as the thermal conductivity of the material. The minus sign means that heat flow is in a direction of decreasing temperature.

$$dQ/dt = -K_T A dT/dx$$

7.5 The internal energy U of an ideal gas containing N molecules is proportional to the product of N and the absolute temperature T. The constant of proportionality is 3k/2 where k is Boltzmann's constant.

$$U = (3/2)NkT$$

7.6 In an idealized heated solid called a cavity radiator, the energy radiated from the cavity interior per unit area (called total cavity radiancy, R_C) is proportional to the fourth power of the absolute temperature T. The constant of proportionality σ is called the Stefan-Boltzmann constant.

$$R_C = \sigma T^4$$

8. Temperatures and Conversions

C^0 is the symbol for degrees Celsius, F^0 is the symbol for degrees Fahrenheit and K^0 means degrees Kelvin (Absolute).

8.1 Water freezes at 0 C^0 at standard atmospheric pressure.

8.2 Water boils at 100 C^0 at standard atmospheric pressure.

8.3 The triple point of water (existing simultaneously as a gas, liquid and solid) occurs at a temperature of 273.16 K^0 and atmospheric pressure of 611.73 Pascals (Newtons per square meter).

8.4 $C^0/100 = (F^0-32)/180$

8.5 $K^0 = C^0 + 273.16$

Electromechanical Laws

1. **Maxwell's Equations**:

1.1 The source of the electric field (**E**) is charge density ρ_q. $\nabla = (\partial/\partial x, \partial/\partial y, \partial/\partial z)$ is the normal vector operator, ($\bullet$) is the normal vector scalar product and ε_0 is a constant called the permittivity of free space. This law is also known as Gauss's law for electricity. The differential form is

$$\nabla \bullet \mathbf{E} = \rho_q/\varepsilon_0$$

The integral form is

$$\varepsilon_0 \oiint \mathbf{E} \bullet \mathbf{n} \, dS = q$$

where $\oiint$ means integration over the closed surface S, **n** is a unit vector normal to S enclosing the charge q.

1.1.1 Maxwell's first equation and may be used to derive Coulomb's law which states that the force between two charges is proportional to the product of the two charges and inversely proportional to the square of the distance between their charge centers. The force is in a direction on a line drawn between the two charges q_1 and q_2 denoted by the unit vector $\mathbf{r_u}$. $K_C = 1/(4\pi\varepsilon_0)$ will be called Coulomb's constant.

$$\mathbf{F} = \mathbf{r_u} K_C q_1 q_2 / r^2$$

giving rise to the electrical potential energy, V between q_1 and q_2

$$V = K_C q_1 q_2 / r$$

If the charges are both positive or both negative, the force is repulsive (like charges repel one another), otherwise the force is attractive (unlike charges attract one another).

1.2 The source of the magnetic field **B** is zero. This is Maxwell's second equation. This also means that magnetic fields always exist in closed loops and magnetic monopoles do not exist. This law is also known as Gauss's law for magnetism. The differential form is

$$\nabla \bullet \mathbf{B} = 0$$

The integral form is

$$\oiint \mathbf{B} \bullet \mathbf{n} dS = 0$$

where $\oiint$ means integration over any closed surface S, **n** is a unit vector perpendicular to the surface, S.

1.3 Ampere's law is also known as Maxwell's third equation. Current density $\mathbf{J}_i$ and/or dynamic electric fields, $\partial \mathbf{E}/\partial t$ give rise to circulating magnetic fields (**B**). μ_0 is known as the permeability constant of free space. The differential form is

$$\nabla \times \mathbf{B} = \mu_0 \mathbf{J}_i + \mu_0 \varepsilon_0 \partial \mathbf{E}/\partial t$$

where $\nabla = (\partial/\partial x, \partial/\partial y, \partial/\partial z)$ is the Del vector operator and **x** is the vector cross product. The integral form is

$(1/\mu_0) \oint \mathbf{B} \bullet \mathbf{ds} = i$

where $\oint$ means integration over a closed line s, circulating around the electrical current, i. **ds** is an infinitesimal vector line element of s, that **B** circulates through. **B** is perpendicular to the direction of the electrical current i.

1.4 Faraday's law is also known as Maxwell's fourth equation. Dynamic magnetic fields, $(\partial \mathbf{B}/\partial t)$ give rise to circulating electric fields (**E**). The differential form is

$\nabla \times \mathbf{E} = -\partial \mathbf{B}/\partial t$

where $\nabla = (\partial/\partial x, \partial/\partial y, \partial/\partial z)$ is the normal Del vector operator and **x** is the vector cross product.

The integral form is

$$\oint \mathbf{E} \bullet \mathbf{ds} = -\iint (\partial \mathbf{B}/\partial t) \bullet \mathbf{n} dS = -\partial \Phi/\partial t$$

where $\Phi = \iint \mathbf{B} \bullet \mathbf{n} dS$ is called the magnetic flux in which **B** penetrates the surface area S. **n** is a unit vector perpendicular to the surface area S.

2. The Lorentz Force:

The force **F** on a charge q moving with velocity **v** by an external electric field **E** and by an external magnetic field **B** and **x** is the normal vector cross product.

$$\mathbf{F} = q\mathbf{E} + q\mathbf{v} \times \mathbf{B}$$

3. Electromagnetic Wave Equations:

When there is no charges or currents, as in the vacuum of matter free space, Maxwell's equations yield a wave equation that is satisfied by both the electric field **E** as well as the magnetic field **B**.

These equations yields the precise description of induced electromagnetic fields.

3.1 $\nabla^2 \mathbf{E} - \partial^2 \mathbf{E}/(c^2 \partial t^2) = 0$ and

3.2 $\nabla^2 \mathbf{B} - \partial^2 \mathbf{B}/(c^2 \partial t^2) = 0$

where $\nabla^2 = \nabla \bullet \nabla = \partial^2/\partial x^2 + \partial^2/\partial y^2 + \partial^2/\partial z^2$, t is the time and c is the speed of light in vacuum. Note that if one utilizes the gradient operator, $\square$ defined as $\square = (\partial/\partial x, \partial/\partial y, \partial/\partial z, \partial/\partial(ict))$ then, the Dalembertian operator, $\square^2 = \square \bullet \square = \partial^2/\partial x^2 + \partial^2/\partial y^2 + \partial^2/\partial z^2 - \partial^2/(c^2 \partial t^2)$ makes the electromagnetic wave equations 3.1 and 3.2 simplify to

3.1.1 $\square^2 \mathbf{E} = 0$ and

3.2.1 $\square^2 \mathbf{B} = 0$

Conservation Laws

1. Conservation of energy: A system's total energy, E_T is the same both before (B) and after (A) any energy transformation.

$$(E_T)_B = (E_T)_A$$

2. Conservation of momentum: A system's total momentum, p_T is the same both before and after any energy transformation.

$$(p_T)_B = (p_T)_A$$

3. Conservation of angular momentum: A system's total angular momentum, L_T is the same both before and after any energy transformation.

$$(L_T)_B = (L_T)_A$$

4. Conservation of charge: A system's total charge, Q_T is the same both before and after any energy transformation.

$(Q_T)_B = (Q_T)_A$

5. Conservation of baryon number: A system's baryon number, N_B is the same both before and after any energy transformation. Baryons are composed of quarks. Quarks have baryon number $+1/3$. Antiquarks have baryon number $-1/3$.

$(N_B)_B = (N_B)_A$

6. Conservation of lepton number: A system's lepton number, N_L is the same both before and after any energy transformation.

$(N_L)_B = (N_L)_A$

7. For any energy system, another related energy system predicted by the simultaneous operations of time reversal, charge conjugation (signs of all

charges involved are reversed) and space reversal (mirror image or parity) is also possible. This is called CPT for short. Below, E_T is the total energy of a system and BCPT means before the CPT operation and ACPT means after the CPT operation.

$(E_T)_{BCPT} = (E_T)_{ACPT}$

Basic Units

position: (measured with a ruler)

meter = m

mass: (measured with a balance scale)

kilogram = kg

time: (measured with a clock)

second = s

charge: (measured with a voltmeter)

coulomb = coul

Equivalent Units

Force: Newton = nt = kg–m/s^2

Pressure: Pascal = nt/m^2

Energy: joule = nt–m

Inductance: henry = joule–m–s^2/coul^2

Capacitance: farad = coul^2/joule

Basic Physical Constants

Name	Symbol	Value
Speed of light	c	3.00×10^8 m/s
Gravitational Constant	G	6.67×10^{-11} nt-m²/kg²
Avogadro's number	N_0	6.023×10^{23} /mole
Universal Gas Constant	R	8.32 joules/(mole–K⁰)
(Planck's constant)/2π	$\hbar$	1.055×10^{-34} joule–s
Planck length	$L_P = (\hbar G/c^3)^{1/2}$	1.616×10^{-35} m
Planck time	$T_P = (\hbar G/c^5)^{1/2}$	5.391×10^{-44} s
Planck mass	$M_P = (\hbar c/G)^{1/2}$	2.177×10^{-8} kg
Boltzmann's constant	k	1.38×10^{-23} joules/(molecule–K⁰)
Stefan-Boltzmann constant	σ	5.67×10^{-8} joules/m²/(K⁰)⁴
Permeability constant	μ_0	1.26×10^{-6} henry/m

Name	Symbol	Value
Permittivity constant	ε_0	1.26×10^{-6} farad/m
Electron charge	q_e	-1.6022×10^{-19} coul
Electron rest mass	m_e	9.11×10^{-31} kg
Proton rest mass	m_p	1.67239×10^{-27} kg
Neutron rest mass	m_N	1.6747×10^{-27} kg
Coulombs constant	$1/(4\pi\varepsilon_0)$	8.99×10^9 nt–m^2/coul^2

Basic Elementary Particles

Preliminary Particle Descriptors

1. Family Names – Particles belong to functional families having a set number of family members. For example, the gluon family has eight members and they function to provide the strong nuclear force that hold quarks together. Individual particles have both a historical name and a symbol. For example, an electron has the symbol e^-.

2. Color – Quarks can either be red, green or blue (r,g,b). Anti-quarks can either be –red, –green or –blue (–r, –g, –b). This is similar to charge coming in two types, the minus (–) and the plus (+) type.

3. Charge – measured in units of positive electronic charge or the charge on a positron (anti–electron). The charge magnitude of a negative electron (e^-) or a positive positron (e^+) are equal. An anti–particle has the opposite charge as the particle.

Basic Elementary Particles   105

4. Spin – Axial angular momentum measured in units of Planck's constant divided by 2π and denoted by $\hbar$. Quantum Spin is specified as positive, but it is understood that quantum mechanically, it can either be positive (parallel) or negative (anti–parallel) to any given direction. Fermions (matter particles) have half integral values of $\hbar$. Bosons (force field particles) have integral values of $\hbar$.

5. Helicity – Helicity is also given in terms of $\hbar$ and may be thought of as the component of the particle's spin in the direction of the particle's velocity vector. The helicity of particles moving at the velocity of light is different than the helicity of particles that do not. Particles moving at the velocity of light, c such as photons, must have zero rest mass and there is no coordinate system for which its velocity is zero. Thus, the component of a photon's spin ($\hbar$) along its velocity vector is the same as its spin orientation, either $+\hbar$ or $-\hbar$ since it cannot be observed at rest. Thus, a photon has an intrinsic helicity the same as

its intrinsic spin. On the other hand, particles with non-zero rest mass have non–intrinsic helicity dependent on the observer since their spin can be observed when they are at rest and their spin components in the direction of motion must have a quantum difference of $+\hbar$. For example, the weakons, responsible for the electroweak forces, with non-zero rest masses and spin of $\hbar$ have helicity of either $-\hbar$, 0, or $+\hbar$. A particle and its anti–particle have opposite helicity.

6. Rest Mass – Measured in either Proton rest masses (Mp) or millions of electron volts (Mev). An electron volt (1.602×10^{-19} joules) is the kinetic energy an electron gains by being propelled a distance of one meter by an electrical field of strength, one volt per meter. The equivalent energy of a proton at rest is 938 Mev. The reason rest mass can be measured in terms of energy is because of Einstein's famous equation $E_0 = m_0 c^2$ which relates rest mass, m_0 to rest mass energy, E_0 by a constant, being the square of the speed of light, c^2.

7. Field Energy – Force fields are caused by corresponding field particles having integral values of $\hbar$ (called bosons). Matter particles having half integral values of $\hbar$ (called fermions) are influenced by force fields caused by their interaction with the corresponding boson. The four force fields are strong nuclear (gluons), electroweak (weakons), electromagnetic (photons) and gravitational (gravitons).

Anti–Particle Properties

All particles have an anti–particle. The anti–particle has the opposite charge of the particle. The anti–particle has the opposite helicity of the particle. The anti–particle of a non–zero rest mass particle having zero charge, and having a spin of one $\hbar$ and zero helicity is the particle itself. The antiparticle has the same mass as the particle. A particle and its anti–particle (that is not itself) annihilate one another upon contact in a burst of other energetic particles.

Matter Energy Particles

All material energy is composed of fundamental matter particles experimentally observed to exist as three energy families (UP, CHARMED, TOP) of four fermions each, in its simplest representation. Two of the fermions are light and are called leptons and two of the fermions are heavy and are called quarks. One of the leptons carries a negative electronic charge, the other has no charge.

Origin of the UP Family

The nuclei of atoms are composed of neutrons and protons. A neutron consists of two (red and blue) down quarks, ($d_R^{-1/3}$, $d_B^{-1/3}$, $u_G^{2/3}$) and one (green) up quark, . A proton consists of two (red and blue) up quarks, and one (green) down quark, ($u_R^{2/3}$, $u_B^{2/3}$, $d_G^{-1/3}$). Any other cyclic permutation of red, green or blue colored quarks in neutrons or protons is possible. The proton is stable. An isolated neutron, n is unstable and will decay into a proton, p electron, e^- and an electron anti–neutrino, $\acute{υ}_e$. The net effect is that one of the down quarks of the neutron will change into an electron, anti–neutrino and an up quark. This effectively transformed the internal structure of a neutron ($d_R^{-1/3}$, $d_B^{-1/3}$, $u_G^{2/3}$) into that of a proton ($u_R^{2/3}$, $u_B^{2/3}$, $d_G^{-1/3}$). The up quark has a charge of 2/3 e^+ while the down quark has a charge of $-1/3$ e^+. Thus a proton has a net charge of e^+ while the neutron has a net charge of 0. The UP family making up neutrons and protons consist of four family members which are the up quark, down quark, electron and its anti–neutrino. There are two other four member families. The TOP family has the highest rest mass energy particle members. The CHARMED family has intermediate rest mass energy particle members. The UP family has the lowest rest mass energy particles. Each family maintains the same relationships between its members.

The UP Family

The UP family consists of an up quark, $u^{2/3}$, a down quark, $d^{-1/3}$, electron, e^-, with its electron anti–neutrino, $ύ_e$. The quarks can either be red, blue or green. The up quark has a charge of 2/3 e$^+$ while the down quark has a charge of −1/3 e$^+$. The electron has a rest mass energy of .511 Mev. These particles have the lowest rest mass energy and represent the ground state rest mass energy of the matter families. All UP fermion family members have a spin of ½$\hbar$ and a helicity of plus or minus ½$\hbar$.

The CHARMED Family

The CHARMED family consists of a charmed quark, c$^{2/3}$, a strange quark, s$^{-1/3}$, muon, μ^- with its muon anti–neutrino, $\acute{\upsilon}_\mu$. The quarks can either be red, blue or green. The charmed quark has a charge of 2/3 e$^+$ while the strange quark has a charge of − 1/3 e$^+$. The muon has a rest mass energy of 105.66 Mev. These particles have intermediate energy and represent a higher rest mass energy state than the UP family. All CHARMED fermion family members have a spin of ½$\hbar$ and a helicity of plus or minus ½$\hbar$.

The TOP Family

The TOP family consists of a top quark, $t^{2/3}$, bottom quark, $b^{-1/3}$, tauon, τ^- with its tauon anti–neutrino, $\acute{\upsilon}_\tau$. The quarks can either be red, blue or green. The top quark has a charge of 2/3 e^+ while the bottom quark has a charge of $-1/3$ e^+. The tauon has a rest mass of 1784.2 Mev. These particles have the highest rest mass energy state and represent a higher energy state than that of the CHARMED family. All TOP fermion family members have a spin of ½$\hbar$ and a helicity of plus or minus ½$\hbar$.

Field Energy Particles

Gluon Family

Gluons ($g_1 - g_8$) are responsible for the strong force field between the three colored (red, green and blue) quarks making up protons and neutrons, of which all nuclei are composed. There are eight different gluons. Gluons carry color combinations (r, g, b, –r, –g, –b) and compose the gluon field holding quark trios together in protons and neutrons. Gluons have a spin of $\hbar$. Gluons have zero rest mass and therefore move at c, the velocity of light. Thus, gluons have helicity of either plus or minus $\hbar$.

Photon Family

Photons (γ) are responsible for the electromagnetic forces which act between charges. Photons have no color and no charge. Photons have a spin of $\hbar$. Photons have no rest mass and move at the velocity of light. Thus, photons have helicity of either plus or minus one $\hbar$. The positive helicity photon is the anti–photon of the negative helicity photon. While in flight, photons have mass, energy and momentum.

The Weakon Family

Weakons give rise to the electroweak force field responsible for radioactive decay. Recall that a neutron is composed of two down quarks and one up quark. The decay of an isolated neutron is an example of radioactive beta (electron) decay in which one of the down quarks in a neutron decays into a weakon (the omega minus) which then decays into an up quark, electron and anti–neutrino. The net effect is that a neutron decays into a proton, electron and anti–neutrino. There are three different weakons, the omega minus (Ω^-), omega zero or zeta (Z^0) and the omega plus (Ω^+). These weakons have no color and carry charges of e^-, 0, e^+ respectively. Weakons have a spin of $\hbar$ and each can have helicity of $-\hbar$, 0 or $+\hbar$. Weakons have rest masses of 85 Mp, 260 Mp and 85 Mp respectively. Anti–weakons have opposite charges and helicities as the corresponding weakons.

The Meson Families

Mesons give rise to the forces between baryons (quark trios). Mesons are not elemental but are composed of quark anti–quark pairs (combos taken from any of the three families of quarks) and are mentioned here for completeness. Obviously, there are many families of mesons, and the pi meson family (pions) are responsible for forces between nucleons (either neutrons or protons). Pions will be presented next as an example.

The Pi Meson (Pion) Family

The Pions are responsible for forces between nucleons (either neutrons or protons) and are composed of quark anti–quark pairs. The pi minus (π^-) is composed of a down quark with a charge of $-1/3$ e^+ and an anti–up quark with a charge of $-2/3$ e^+ for a total charge of e^-. The pi zero (π^0) is a mixture of an up quark and an anti-up quark, with a down quark and an anti–down quark. The pi plus (π^+) is composed of an up quark with a charge of $+2/3$ e^+ and an anti–down quark with a charge of $+1/3$ e^+ for a total charge of e^+. These pions have no color and carry charges of e^-, 0, e^+ respectively. Pions have a spin of 0 and each has helicity of 0. The charged pions have rest masses of 139.57 Mev, while the pi zero has a rest mass of 134.96 Mev. The anti–pi minus is the pi plus. The anti–pi plus is the pi minus. The anti pi zero is the pi zero itself.

Graviton Family

Gravitons (G_- and G_+) are responsible for the gravitational force fields which act between masses. Gravitons have no color and no charge.

The G_- graviton has a spin of $2\hbar$ and a zero rest mass. It moves at the velocity of light and thus, its helicity is $-2\hbar$ or $+2\hbar$. It is assumed to have negative mass in flight while being exchanged between any two positive masses or any two negative anti–masses. This is because the gravitational potential energy between two positive masses or two negative masses is negative.

Because of a new scientific theory called "Nature of the First Cause", in which positive matter is gravitationally repelled by negative anti–matter, the G_+ graviton is postulated to exist. It also has a spin of $2\hbar$. It is assumed to have a zero rest mass and moves with the velocity of light and has positive mass in its flight between negative antimatter and positive matter. Thus, the G_+ graviton also has helicity of $-2\hbar$, or $2\hbar$. The G_+ gravitons fill up all space and are responsible for the force of repulsion between negative anti–matter and positive matter. By the "First Cause" theory, it makes up the repulsive gravitational field which is responsible for the accelerated expansion of distant positive matter in the universe (galaxies not in the local group).

The Higgs Family

There are two Higgs bosons (H_L and H_H) called the light Higgs boson, H_L of the unified electroweak theory and the heavy Higgs boson, H_H of the grand unified theory. The heavy Higgs boson, makes up the Higgs field and permeates all space. This field is responsible for assigning masses to all the fundamental particles. The light Higgs boson is responsible for assigning the masses to the weakons. Both Higgs bosons have a spin of zero (0), and thus they both have a helicity of zero. Both Higgs bosons have non–zero rest masses with the light Higgs rest mass at roughly 10^5 Mev and the heavy Higgs rest mass of about 10^{17} Mev.

Complete Set of Particles

All matter particles which have been discovered are combinations of the above elementary matter particles. All the known force fields consists of varying energy and intensity of the above force field particles.

The Hadrons (consisting of quarks) which are matter particles that have been discovered now number over two hundred which exceed the number of known elements.

References

Al-Khalili, Jim, *Quantum, A Guide for the Perplexed*, United Kingdom, Weidenfeld & Nicolson, 2003

Ames, Joseph Sweetman & Murnaghan, Francis D., *Theoretical Mechanics An Introduction to Mathematical Physics*, New York, Dover Publications, Inc., 1957

Atkins, K. R., *Physics*, New York, John Wiley & Sons, Inc., 1965

Bergmann, Peter Gabriel, *Introduction to the Theory Of Relativity*, New York, Dover Publications, Inc., 1976

Blass, Gerhard A., *Theoretical Physics*, New York, Appleton-Century-Crofts, 1962

Born, Max, *Einstein's Theory of Relativity*, New York, Dover Publications, Inc., 1962

Breithaupt, Jim, *Cosmology*, Blacklick, OH, McGraw-Hill, 1999

Davies, Paul, *The New Physics*, New York, Cambridge University Press, 1996

De Broglie, Louis, *matter and light*, New York, Dover Publications, Inc., 1939

Einstein, Albert, *Builders of the Universe*, Los Angeles, CA, U. S. Library Association, Inc., 1932

Einstein, Albert, & Lorentz, H. & A., Minkowski, H., & Weyl, H., *The Principle of Relativity*, New York, Dover Publications, Inc., 1952

Einstein, Albert, *Relativity The Special and General Theory*, New York, Crown Publishers, Inc., 1961

Feynman, Richard P., *QED The Strange Theory of Light and Matter*, Princeton, New Jersey, Princeton University Press, 1988

Feynman, Richard P., *Six Not So Easy Pieces*, New York, Basic Books, 1997

Frankel, Theodore, *Gravitational Curvature An Introduction to Einstein's Theory*, San Francisco, W. H. Freeman and Company, 1979

Goldstein, Herbert, *Classical Mechanics*, London, Addison-Wesley Publishing Company, Inc., 1950

Halliday, David & Resnick, Robert, *Physics For Students of Science and Engineering*, New York, John Wiley & Sons, Inc., 1962

Hawking, Stephen & Penrose, Roger, *The Nature of Space and Time*, New Jersey, Princeton University Press, 1996

Heisenberg, Werner Karl, *The Nature of Elementary Particles*, in Physics Today, Page 39, March 1976

Kaku, Michio, *Hyperspace*, New York, Anchor Books, 1995

Kaku, Michio, *Parallel Worlds*, New York, Anchor Books, 2006

Kaplan, Irving, *Nuclear Physics*, Reading, Massachusetts, Addison-Wesley Publishing Company, Inc., 1962

McMahon, David, *quantum mechanics demystified*, New York, McGraw Hill, 2005

Messiah, Albert, *Quantum Mechanics*, New York, Dover Publications, Inc., 1999

Park, David, *Introduction to the Quantum Theory*, New York, McGraw-Hill Book Company, 1964

Penrose, Roger, *The Road To Reality, A Complete Guide to the Laws of the Universe*, New York, Vintage Books, 2004

Powell, John L. & Crasemann, Bernd, *Quantum Mechanics*, Reading, Massachusetts, Addison-Wesley Publishing Company, Inc., 1961

Ridpath, Ian, *The Illustrated Encyclopedia of the Universe*, New York, Watson-Guptil Publications, 2001

Sears, Francis W., & Zemansky, Mark W. & Young, Hugh D., *College Physics*, Menlo Park, California, Addison-Wesley Publishing Company, 1986

Segre, Emilio, *Nuclei and Particles*, New York, W. A. Benjamin, Inc., 1964

Shortley, George & Williams, Dudley, *Elements of Physics For Students of Science and Engineering*,

Englewood Cliffs, New Jersey, Prentice-Hall, Inc., 1965

Van Heuvelen, Alan, *Physics, A General Introduction*, Boston, Little, Brown and Company, 1982

Weinberg, Steve, *Dreams of a Final Theory: The Search for the Fundamental Laws of Nature*, New York, Pantheon Books, 1992

Weyl, Hermann, *Symmetry*, Princeton University Press, 1952

Young, Hugh D., *Statistical Treatment of Experimental Data*, McGraw-Hill Co., Inc., 1962

INDEX

A

absolute temperature, 89, 90

acceleration, 75, 77, 80

Acceleration, 74

action, vii, 24, 70, 80

Action, xvii, 24, 70

angular frequency, 33, 84

angular momentum, 67, 78, 82

Angular momentum, 77

anti–neutrino, 109, 110, 111, 113

anti–particle, 104, 106, 107

Anti–Particle, 107

Anti–weakons, 113

average, 74, 75, 76

Axial angular momentum, 105

B

basic units, xiv

big bang, vii

Big Bang, iv, xvi

Boltzmann's constant, 90, 102

Born Interpretation, 53

Bosons, 105

bottom quark, 111

C

causality, 42, 65, 69

cavity radiancy, 90

Celsius, 91

charge, 78, 92, 93, 96, 101, 103

Charge, 104

charge conjugation, 99

Charge density, 78

charges, 93, 96, 100

CHARMED family, 109, 110, 111

charmed quark, 110

Color, 104

commute, 67

Commute, 70

commutes, 60, 67

complex conjugate, 53, 54, 56, 57

conductivity, 79, 89
Conservation, xviii, xix, xx, 53, 54, 59, 61
Conservation Laws, 98
Conservation of angular momentum, 98
Conservation of baryon number, 99
Conservation of charge, 99
Conservation of energy, 98
Conservation of lepton number, 99
Conservation of momentum, 98
constants of the motion, 67
Coulomb, 93
Coulomb's law, 93
creation, vii
current, 78, 79, 95

D

Dalembertian operator, 5, 97
De Broglie, 85
De Broglie's, 85
Dirac, xii, xiii, xvi, 1, 2, 9, 10, 11, 12, 15, 64, 70
down quark, 108, 109, 114

E

eigenvalue, 3, 82

Eigenvalue, 70

eigenvalues, 67, 82

Einstein, 85, 119, 120

Einsteinian, 73, 74, 75, 76, 77, 82, 87

Einstein's, 85

electric field, 79, 92, 96

electric fields, 95

Electrical current, 78

electromagnetic, 84, 107, 112

Electromagnetic, 96

electron, 104, 106, 108, 109, 113

Electron, 103

electron anti–neutrino, 108, 109

electron volt, 106

electrons, 1, 2, 11, 22

electroweak, 106, 107, 112, 116

Elementary Particles, 104

energy, ix, 73, 74, 75, 76, 77, 81, 82, 83, 84, 87, 88, 90, 93, 98, 99

   internal, 88, 90

kinetic, 87
potential, 81, 93
total, 76, 82, 98
transformation, 98
Energy, 82, 101
evolution, 42, 47, 51, 52, 65, 68, 69
Evolution, xviii, 46, 68
evolves, 45
expectation value, 56
experiment, 83

F

Fahrenheit, 91
Family Names, 104
fermions, 107, 108
Fermions, 105
Field Energy, 107, 111
field of force, 32
field particles, 107, 117
fields, 94, 95
electric, 79, 92, 94, 96
gravitational, 81
magnetic, 93

first order, 42, 87

force field particles, 105

force fields, xii, 14, 107, 115, 117

four vector, 23, 24

free particle, xiii, 6, 14, 15, 16, 25

Free Particle, xvi, xvii, 6, 14, 26, 28

free relativistic particle, 30, 34

free system at rest, 7

fundamental physical laws, xiv

G

galaxies, 116

Gauss, 92, 94

generalize, 31, 39

generalized, xiii, 17, 27, 29, 36, 37, 40

gluons, 107, 111

Gluons, 111

gradient operator, 4, 97

gravitational, 81, 82, 107, 115, 116

Gravitational, 81, 102

gravitational field, viii

graviton, 115

Graviton Family, 115

gravitons, 9, 12, 33, 107, 115
Gravitons, 115

H

Hamiltonian, xvi, xvii, xviii, xix, 3, 4, 20, 30, 34, 42, 43, 44, 45, 59, 66, 67, 82
hardware, vi
heavy Higgs, 116
Heisenberg, 83, 84
Heisenberg uncertainty principle, 84
Helicity, 105
Higgs bosons, 116
Higgs Family, 116

I

infinitesimal volume, 53, 82
instantaneous, 74, 75, 76, 78, 89
intrinsic helicity, 105
Invariant, xvii, 30, 61, 71
invariant form, xiii, 30, 41
irreversibility, 89

K

Kelvin, 91

kinetic, 4, 8, 12, 14, 15, 20, 21, 26, 29, 34, 38, 43, 44, 66, 87

kinetic energy, 87

Kinetic Energy Operator, xviii, xix, 38

Klein-Gordon, xii, xiii, xv, xvi, 1, 2, 4, 5, 6, 8, 10, 11, 12, 39, 41, 63, 64, 71

Klein–Gordon, xii, 1, 2, 15

L

length, 86, 102

leptons, 108

light Higgs, 116

local group, 116

Lorentz, xii, 1, 2, 9, 30, 42, 61, 62, 63, 70, 71, 96, 119

Lorentz Force, 96

Lorentz invariant, xii, 1, 2, 9, 42, 62, 63, 70, 71

M

magnetic field, 93, 96

magnetic fields, 94

magnetic moment, 12

mass, 75, 77, 79, 80, 81, 85, 87, 101, 102, 103

Mass, 77, 81, 82

material energy, 108

matter, vii, 119

matter particles, 105, 108, 117

Maxwell's, 92, 93, 94, 95, 96

Maxwell's equations, 96

measurement, 83, 84

Meson Families, 113

Mesons, 113

momentum, 6, 7, 14, 15, 23, 24, 27, 28, 67, 70, 75, 76, 78, 82, 83, 84, 85

Momentum Operator, xvii, 27

motion, xiii, 4, 62, 67, 80, 87

muon, 110

N

negative anti–matter, 115

neutrinos, 9, 12, 33

neutron, 108, 112

Neutron, 103

newton, 101

Newtonian, 73, 74, 75, 76, 81, 87, 88

Newtonian dynamics, 13

Newtons, 91

Newton's, 80, 81

non-relativistic, xiii, 12, 22, 67, 72

non-zero rest mass, xiii

Normalization, xix, 60

normalization integral, 61

O

omega minus, 113

omega plus, 113

operator, 3, 4, 20, 21, 27, 29, 30, 34, 38, 43, 44, 45, 56, 59, 60, 67, 82, 92, 95

operator notation, 21

operators, 29, 66, 82

Operators, xvii, 29

origin, 16, 77

P

pair, viii

particle, xiii, 2, 6, 15, 25, 32, 48, 50, 51, 53, 68, 69, 70, 85

particles, viii, xii, xiii, 9, 51

Pascal, 101

Pascals, 91

philosophy, xiii, 8, 67

photon, 84

photons, viii, 9, 12, 33, 105, 107, 112

Photons, 112

physical constants, xiv

Pions, 113, 114

Planck length, ix

Planck Length, viii

Planck mass, ix

Planck Mass, viii

Planck time, ix

Planck Time, viii

Planck's constant, 3, 83, 84, 85, 102

Planck's Constant, 71

position, 73, 74, 77, 83, 84

Position, 73, 101

positive matter, 115

positron, 104

positrons, 11

potential, xii, xiii, 4, 8, 9, 12, 14, 17, 21, 32, 33, 36, 40, 41, 43, 52, 60, 62, 67, 68, 81, 93

potential energy, xii, xiii, 4, 8, 9, 12, 14, 17, 21, 32, 33, 36, 40, 41, 43, 52, 60, 67, 68

pressure, 77, 88, 89, 91

Pressure, 77

probability, 53, 54, 55, 56, 61, 82, 83

Probability, xviii, 53, 54

probability current, 55, 56

Probability Current, xviii, 54

proton, 106, 108, 113

Proton, 103

protons, 1

Q

quantum, ix, 1, 50, 61, 65

quantum mechanics, vi

quark anti–quark pairs, 113

quarks, 99, 104, 108, 109, 110, 111, 112, 113, 114, 117

R

relativistic equations, 67

relativistic wave equation, xi, xii, xiii, 1, 2, 12, 22, 39, 70, 71, 72

rest, ix, xii, xiii, 2, 3, 7, 8, 9, 11, 12, 15, 24, 27, 28, 33, 48, 50, 62, 67, 68, 77, 80, 86, 87, 103

Rest Mass, 106

rest mass energy, 3, 7, 24

reversibility, 89

S

scalar product, 24, 92

Schrödinger, xiii, xvi, 1, 12, 13, 14, 15, 16, 18, 20, 22, 63, 64, 67, 72

second order, 65

solution, viii

Solution, xvi, 32

space reversal, 100

Spacetime, xvi

speed of light, 2, 13, 72, 73, 97

Speed of Light, 72

spin, 67, 82

Spin, 105

spinor, 11

standard deviation, 83, 84

stationary energy states, xii, 68

stationary free particles, 15

stationary states, 13, 19, 33, 64

Statistical, xix, 59

Stefan-Boltzmann constant, 90

strange quark, 110

strong nuclear, 104, 107

T

tauon, 111

thermodynamics, 88

three dimensional position, 6, 24

three dimensions, 17, 27, 35, 37

time, 73, 74, 76, 78, 86, 97, 101, 102

time dependence, xii, 18, 19

time dependent, xii, 7, 13, 20, 21, 26, 34, 40, 42, 44, 45, 46, 63, 64

Time Dependent, xvi, xviii, xix, 35, 40, 63

time independent, 13, 18, 21, 22, 34, 39, 43, 45, 46, 50, 62, 63, 64, 68

Time Independent, xvi, xvii, xix, 17, 28, 32, 62, 63
time independent equation, 13, 18, 21, 22, 34, 43, 45, 46, 63
time reversal, 99
TOP family, 109, 111
top quark, 111
total dynamic energy, 41
total energy, 3, 6, 8, 18, 24, 32, 42, 52, 58, 60, 68, 82, 98, 100
transformation, 71, 98, 99
triple point of water, 91

U

uncertainty principle, 84
Uncertainty Principle, xv, 83
UP family, 109, 110
UP Family, 108, 109
up quark, 108, 109, 112, 114

V

vector, 74, 75, 77, 79, 81, 92, 93, 94, 95, 96
vector cross product, 78, 95

velocity, 7, 14, 24, 51, 62, 63, 69, 71, 74, 75, 76, 77, 80, 85, 86, 87, 96

Velocity, 74

velocity of light, 63, 85, 86

W

Water boils, 91

Water freezes, 91

wave amplitude, 15, 26, 27

wave equation, xiii, 96

Wave Equation, xvi, xvii, 28, 82

   electromagnetic, 96

   quantum mechanical, 82

wave equations, 1, 42, 58, 97

wave function, xiii, 2, 4, 6, 7, 8, 11, 14, 15, 16, 24, 25, 26, 27, 35, 42, 45, 47, 48, 49, 50, 51, 52, 53, 64, 65, 67, 68, 69, 70, 82, 83

Wave Function, xviii, 6, 14, 46, 48, 51

weakon, 113

weakons, 106, 107, 113, 116

Weakons, 112

Z

zero rest mass, xii, xiii, 7, 9, 12, 15, 33, 62

zeta, 113

www.ingramcontent.com/pod-product-compliance
Lightning Source LLC
Chambersburg PA
CBHW051524170526
45165CB00002B/591